AF477989

The Institute of Mathematics
and its Applications
Conference Series

Previous volumes in this series were published by
Academic Press to whom all enquiries should be addressed.
The following and all forthcoming volumes will be published by
Oxford University Press throughout the world.

Continued overleaf

Mathematics of Dependable Systems

Based on the proceedings of a conference on Mathematics of Dependable
Systems, organized by The Institute of Mathematics and its Applications and held
at Royal Holloway, University of London, in September 1993.

Edited by

CHRIS MITCHELL

and

VICTORIA STAVRIDOU
Royal Holloway, University of London

CLARENDON PRESS · OXFORD · 1995

Oxford University Press, Walton Street, Oxford OX2 6DP

Oxford New York
Athens Auckland Bangkok Bombay
Calcutta Cape Town Dar es Salaam Delhi
Florence Hong Kong Istanbul Karachi
Kuala Lumpur Madras Madrid Melbourne
Mexico City Nairobi Paris Singapore
Taipei Tokyo Toronto

and associated companies in
Berlin Ibadan

Oxford is a trade mark of Oxford University Press

Published in the United States
by Oxford University Press Inc., New York

A catalogue record for this book is available from the British Library

Library of Congress Cataloging in Publication Data

ISBN 0 19 853491 4

Printed in Great Britain by
Bookcraft (Bath) Ltd
Midsomer Norton, Avon

PREFACE

The last ten years have brought a very rapid growth in computerised control in all areas of technology. In the past software and hardware design errors were relatively unimportant, costing only time and modest sums of money. However, the introduction of automatic control for systems on which human lives depend, e.g. airliner, manufacturing plant, or nuclear power station control sytems, has greatly increased the requirement to eliminate all types of failure, whether they derive from design errors, operational problems or sabotage. The need for the dependable operation of automatic systems is at the heart of the growing subject of *Dependable Systems*.

This area of Computer Science embraces a variety of disciplines which have largely developed independently of one another, including:

- *Security*, directed at protecting systems against unauthorised access and/or modification.

- *Reliability*, the goal of which is to mathematically and statistically model the error performance of large systems.

- *Availability*, covering the problems of providing uninterrupted access to computer and communications systems.

- *Safety*, directed at problems arising in particularly critical computing applications, e.g. control in aviation, shipping, power generation and chemical plants.

All of these areas embrace the use of *Formal Methods*, a large area of Computing directed at developing mathematical techniques with the machinery to specify properties of computer systems, and the means for mathematically proving properties about these systems. There are other clear existing links between these different strands of the subject, particularly within the field of pragmatic codes of practice.

The main goal of the First IMA Conference on "Mathematics of Dependable Systems", held at Royal Holloway, University of London, in September 1993, was to bring together researchers in all of these areas. The need for an exchange of ideas is paramount; in the last few years many similar ideas have been pursued independently in parallel areas of research. For example, the development of Evaluation Criteria and 'Security Levels' for Security Systems, starting with the US DoD Orange Book in the mid 1980s, has been parallelled by recent efforts to create levels of assurance for safety-critical software. The focus of the conference was the mathematical methods used by all branches of the subject, and the conference showed how a wide variety of different mathematical methods have been brought to bear on problems of dependability by various parts of the research community.

The collection of papers in this volume, which contains 15 of the 20 papers presented at the meeting, covers a wide diversity of mathematical and application topics, with System Dependability as the common theme. It is hoped that these papers will expand the horizons of many researchers to areas outside of their own specialised corner of the subject. We confidently expect the next conference on this subject, to be held in York in September 1995, will be an even greater success.

C.J. Mitchell and V. Stavridou
Royal Holloway, University of London, Surrey

ACKNOWLEDGEMENTS

The Institute thanks the authors of the papers, the editors, Professor C.J. Mitchell (Royal Holloway, University of London, Surrey) and Dr. V. Stavridou (Royal Holloway, University of London, Surrey) and also Miss Debbie Brown and Miss Adèle Sharples for preparing and typing the papers.

CONTENTS

x

CONTRIBUTORS

T. ANDERSON; Department of Computing Science and BAe Dependable Computing Systems Centre, University of Newcastle upon Tyne, Newcastle upon Tyne, NE1 7RU.

A.W. ANDREWS; 57 Montague Street, Edinburgh, Scotland, EH8 9QS.

N. BIRD; Lucas Engineering and Systems Limited, P.O. Box 52, Shirley, Solihull, West Midlands, B90 4JJ.

J.P. BOWEN; Programming Research Group, Computing Laboratory, University Of Oxford, Wolfson Building, Parks Road, Oxford, OX1 3QD.

T. COCKRAM; Rolls Royce, GP 1-4, P.O. Box 3, Filton, Bristol, BS12 7QE.

R.J. COLE; Software Measurement Laboratory, Glasgow Caledonian University, Cowcaddens Road, Glasgow, Scotland, G4 0BA.

G.V. CONROY; Department of Computation, University of Manchester Institute of Science and Technology, P.O. Box 88, Sackville Street, Manchester.

S.K. DAS; Artificial Intelligence Group, Department of Computer Science, Queen Mary and Westfield College, University of London, Mile End Road, London, E1 4NS.

D. GOLLMANN; Department of Computer Science, Royal Holloway, University of London, Egham, Surrey, TW20 0EX.

R.W.S. HALE; SRI International Cambridge Research Centre, Suite 23, Millers Yard, Mill Lane, Cambridge, CB2 1RQ.

P.A.V. HALL; Department of Computing, The Open University, Walton Hall, Milton Keynes, MK7 6AA.

J.M.J. HERBERT; SRI International Cambridge Research Centre, Suite 23, Millers Yard, Mill Lane, Cambridge, CB2 1RQ.

M. INGLEBY; Safety Critical Systems Unit, British Rail Research, London Road, Derby, and School of Computing and Mathematics, University of Huddersfield, Queensgate, Huddersfield, HD1 3DH.

H. JIFENG; Programming Research Group, Computing Laboratory, University of Oxford, Wolfson Building, Parks Road, Oxford, OX1 3QD.

M. JOSEPH; Department of Computer Science, University of Warwick, Coventry, CV4 7AL.

R. de LEMOS; Department of Computing Science and Department of Chemical and Process Engineering, University of Newcastle upon Tyne, Newcastle upon Tyne, NE1 7RU.

Z. LIU; Department of Mathematics and Computer Science, University of Leicester, University Road, Leicester, LE1 7RH.

J.H.R. MAY; Department of Computing, The Open University, Walton Hall, Milton Keynes, MK7 6AA.

P. MUKHERJEE; School of Computer Science, Royal Holloway, University of Birmingham, Edgbaston, Birmingham, B15 2TT.

J. NORDAHL; Department of Computer Science, Technical University of Denmark, Building 344, DK-2800 Lyngby, Denmark.

D.L. PARNAS; Communications Research Laboratory, Department of Electrical and Electronic Engineering, McMaster University, Hamilton, Ontario, Canada, L8S 4K1.

D.J. PAVEY; Nuclear Systems Branch, Nuclear Electric plc., Barnett Way, Barnwood, Gloucester, GL4 7RS.

C.P. PFLEEGER; Trusted Information Systems (UK) Limited, 41 Surbiton Road, Kingston-upon-Thames, KT1 2HG.

C. PULLEY; Department of Computation, University of Manchester Institute of Science and Technology, P.O. Box 88, Sackville Street, Manchester.

A. SAEED; BAe Dependable Computing Systems Centre, University of Newcastle upon Tyne, Newcastle upon Tyne, NE1 7RU.

E.V. SØRENSON; Department of Computer Science, Technical University of Denmark, Building 344, DK-2800 Lyngby, Denmark.

B.A. WICHMANN; National Physical Laboratory, Teddington, Middlesex, TW11 0LW.

L.A. WINSBORROW; Nuclear Systems Branch, Nuclear Electric plc., Barnett Way, Barnwood, Gloucester, GL4 7RS.

H. ZHU; Department of Computing, The Open University, Walton Hall, Milton Keynes, MK7 6AA.

Simulated data experiment to test a software reliability growth model based on exercise frequencies

A.W. Andrews and R.J. Cole

Software Measurement Laboratory, Glasgow Caledonian University

1 Introduction

This paper is concerned with the development of a new model of software reliability and the experimental testing of the model using simulated data.

Software products are increasing in size, complexity and application area. Software has penetrated all areas of business and scientific applications and impacts the public through the ever-increasing number and power of products and services. Software is used in safety critical areas such as fly-by-wire control of aircraft, and is inevitably finding its way into applications involving mass applications where safety is paramount, in for example, the control of vehicle active suspension systems.

Software reliability models are used to predict the reliability of programs from *failure data*. A number of models take as input the failure data obtained during program execution in the testing phase or in the field. The modelling literature is extensive and reviews may be found in [1] and [5]. A feature of the set of software reliability models is that no *single* model is able to predict accurately the future reliability of a software program from its past failure data.

This is perhaps not so surprising. In this class of models the program is regarded as a 'black box'. Nothing is known or assumed about the internal operation of the program. Only inter-failure times are available as data for estimating the parameters of the model. The failure process is illustrated in Figure 1.

In this paper we investigate the performance of a new reliability model that depends upon information about program *usage*. The program is no longer regarded as a 'black box' and the new model uses the additional information as data. The proposed model requires some knowledge of the running characteristics of the program, the 'program exercise frequency' in addition to the inter-failure times. Its reliability predictions will be compared with two well-known models. They are the early standard model of Jelinski-Moranda [6] and the later standard Littlewood Stochastic Relia-

1

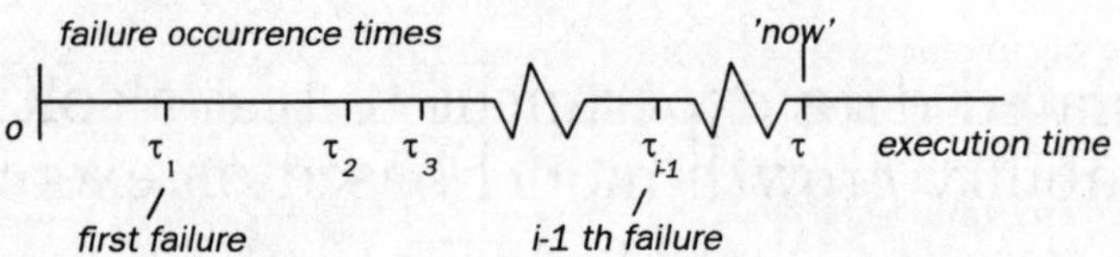

Figure 1. Failure history of a program

bility Growth Model [7]. Key characteristics of these established models will be set out in Section 2 to develop baselines from which to compare the exercise frequency model and its predictions.

2 Basic models of software reliability

2.1 Basic assumptions

The proposed exercise frequency model will be considered and compared with the Jelinski-Moranda and the Littlewood models.

Basic assumptions that are satisfied by all three models considered here are:

Assumption 1 At time 0, the program contains N (unknown) *faults*, or *bugs*.

Assumption 2 When a *failure* occurs, which is the *manifestation* of a fault, the fault is repaired successfully and immediately.

Assumption 3 The failure process of each fault is *independent* of all other faults. A bug will be manifest after a time t with probability density function following an exponential distribution,

$$pdf\,(t) = \phi \exp\,(-\phi t), \tag{2.1}$$

where ϕ is the *occurrence rate* of the bug.

2.2 Bug occurrence rates and the program failure rate

The occurrence rate of a bug is the rate at which failures attributable to that bug would occur if the bug were not removed from the program [4]. From (2.1), the Mean Time To Occurrence (MTTO) attributable to that bug is given by

$$\text{MTTO} = 1/\phi, \tag{2.2}$$

so that the occurrence rate is the reciprocal of the MTTO.

Now consider the program running from time 0 up to τ. Each bug fails as an independent Poisson process with its own occurrence rate. The whole program fails as a Poisson process with failure rate λ being the sum of the occurrence rates of the bugs contained in the program at that time. This follows from the result that a set of superimposed independent homogeneous Poisson processes is itself a Poisson process with rate equal to the sum of the individual rates.

When total execution time is τ and n faults have been fixed with respective occurrence rates ϕ_1, ϕ_2, $\phi_3, \cdots, \phi_n$, the failure process of the *program* leads to a probability density function for time between failures in the form

$$\text{pdf}\,(t) = \lambda_n \exp\left(-\lambda_n t\right), \tag{2.3}$$

where

$$\lambda_n = \phi_{n+1} + \phi_{n+2} + \phi_{n+3} + \cdots, +\phi_N \tag{2.4}$$

is the sum of the occurrence rates of the remaining bugs in the program.

The central interest in the paper is the estimation and comparison of the *program failure rate* λ_n for the three models under consideration.

2.3 The Jelinski-Moranda model

The key J-M assumption is that the occurrence rate of every bug is the same fixed constant. Thus

$$\phi_i = \oslash, \text{ a constant}, \tag{2.5}$$

and the program failure rate after n failures (that is between the n'th and $n+1'$th failure) is given by

$$\lambda_n = (N - n)\oslash \tag{2.6}$$

The model is specified by (2.3) and (2.6) with the unknown parameters N and ϕ being estimated by the maximum liklihood method. The assumption (2.5) of equal constant bug occurrence rates implies that the program failure rate decreases by equal amounts between successive failures, illustrated in Figure 2.

A more revealing figure is to take a plot of program failure rate against failure *number* shown in Figure 3.

The straight line equation corresponds to (2.6), plotting λ_n against n. The gradient of the line is a constant $N\oslash$ involving the fixed but initially

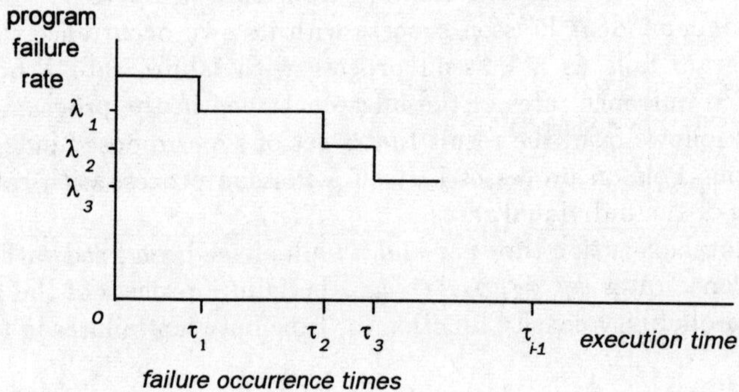

Figure 2. J-M program failure rate against execution time

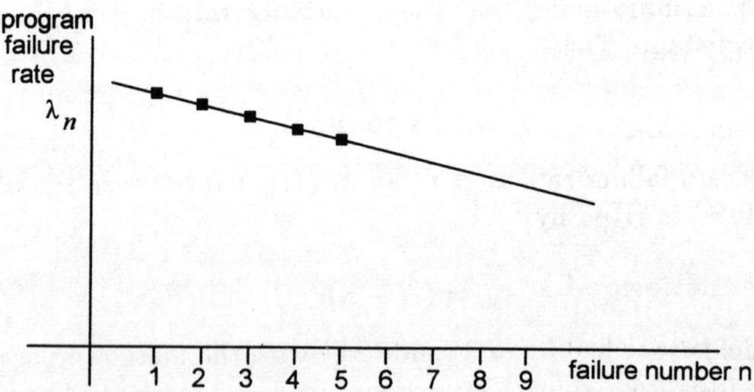

Figure 3. J-M program failure rate against failure number

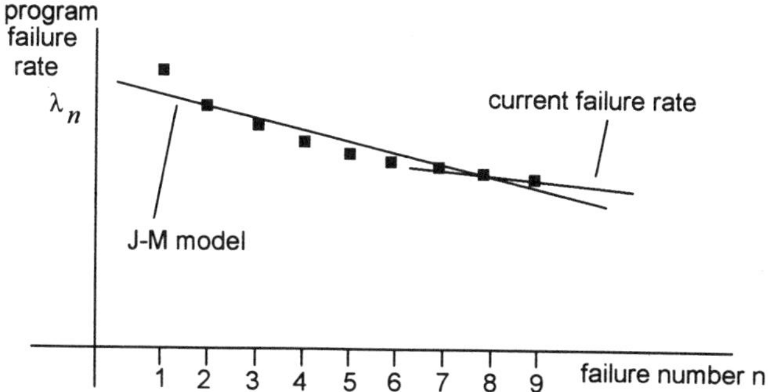

Figure 4. J-M model against realistic program failure rate

unknown number of bugs N and constant but unknown bug occurrence rate $\oslash$. In practice the J-M model turns out to be *optimistic* in respect of the program failure rate and reliability. It has been found [2] that individual bug occurrence rates vary by *several orders of magnitude*. Bugs with higher occurrence rates are more likey to be detected *earlier*, with a larger drop in program failure rate. Assuming perfect and instantaneous bug repair, later bugs will be more likely to have *lower* occurrence rates.

In Figure 4, a more realistic program failure rate sequence is shown against failure number with the J-M straight line giving an optimistic gradient at the end of the fauliure number sequence.

It can be seen that at a given time following n failures, a straight line fit of type (2.6) will underestimate N. Early bugs gives an estimate for the bug occurrrence rate $\oslash$ which is too high with respect to later bugs. The current program failure rate after n failures is estimated assuming that the later bugs to come share the *same* occurrence rate. For these reasons the J-M model does not predict program failure rates well.

2.4 The Littlewood stochastic reliability growth model

An important model, the Littlewood Stochastic Reliability Growth model LSRG was proposed [7], in which the bug occurrence rates are *not* assumed equal, but drawn from a probability distribution. The LSRG model takes into account the variability of bug occurrrence rates. Equations (2.3) and (2.4) still hold but now each occurrence rate is treated as a random variable Φ with some distribution. The program failure rate after n failures Λ_n is a r.v.

$$\Lambda_n = \Phi_{n+1} + \Phi_{n+2} + \Phi_{n+3} + \cdots, + \Phi_N. \tag{2.7}$$

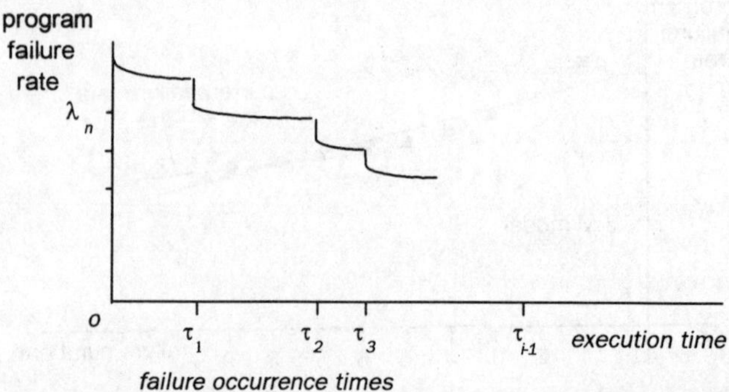

Figure 5. LSRG program failure rate against execution time

The failure process of the program satisfies

$$\mathrm{pdf}\,(t|\Lambda_n = \lambda_n) = \lambda_n \, \exp\,(-\lambda_n t), \qquad (2.8)$$

where λ_n is now a realisation of Λ_n.

The LSRG model led to an analytic result for the program failure rate immediately following n failures in the form

$$\lambda_n = (N - n)/\alpha(\beta + \tau), \qquad (2.9)$$

where α and β are parameters of the distribution of the initial bug occurrence r.v. Φ. The pdf of ϕ is assumed gamma in form and listed in Appendix A along with all other notation. It is worth noting that, as a result of the mixing of different bug occurrence rates, the inter-failure times no longer follow an exponention distribution; the time to failure Xj of the j'th remaining fault has a Pareto distribution in which

$$\mathrm{pdf}\,(x) = \int \phi_j \, \exp\,(-\phi_j x_j) \, \mathrm{pdf}\,(\phi_j)d\phi_j. \qquad (2.10)$$

Result (2.9) is illustrated in Figure 5. It possesses the desired characteristic that early bug fixes cause larger jumps in the program failure rate than later ones. The result also implies a decreasing program failure rate *between* bug fixes as a consequence of the model approach.

Whilst the LSRG model is an enormous improvement on the J-M model and helped surmount its limitations it has been found that the accuracy of its reliability predictions vary [1]. The LSRG model shares with the J-M model the limitation that the program is still regarded as a *black box*, and the only data available which is used to estimate the parameters of the model are the failure times of the program. If more information was available about the running of the program, then a new model could utilise this additional data for its reliability predictions. This new model uses the concept of *exercise frequencies.*

3 A model based on exercise frequencies

Existing software reliability models utilise the occurrence *times* of the first n occurring bugs. If the occurrence *rates* of the first n occurring bugs were also known, then this additional data could be used in a new model to predict program reliability. These rates are not observable directly in practice, but it might be possible to obtain indirect information about the occurrence rates from the *frequency* with which different parts of the program are *exercised.* The concept of exercise frequency provides the basis for the new model presented here.

3.1 The concept of exercise frequency

The starting assumption of the new model is that bugs in frequently exercised parts of the code tend to have higher occurrence rates than bugs in less frequently exercised code. Large software systems are often executed in a non-uniform manner.

A program will be thought of as a structure composed of *program units.* Program *modules* and *statements* are examples of units selected at different levels of decomposition. The program units associated with a software bug are defined to be those which require alteration to remove the bug. The *occurrence rate* of the bug is assumed to be dependent on the *frequency* with which the program units associated with the bug are executed. A bug is 'exercised' whenever one of its associated program units is executed. By monitoring program execution it is possible to observe the exercise frequencies of different program units.

3.2 Linking occurrence rates to exercise frequencies

The program failure rate Λ_n, at time τ 'now', after n failures (2.7) depends upon knowledge of the occurrence rates of the remaining bugs, Φ_{n+1}, $\Phi_{n+2}, \cdots, \Phi_N$, which are unknown. Inferences can be drawn about the distribution of these remaining bugs from the data about the occurrence rates of the n bugs already removed $\Phi_1, \Phi_2, \Phi_3, \cdots, , \cdot\Phi_n$. Direct observation of these rates is not realistically possible in practice. If however a relationship is modelled between bug occurrence rates and their respective exercise frequencies $f_1, f_2, f_3, \cdots, , \cdot f_n$, then the data on failures and exercise frequencies observed through monitoring program execution can be used to predict reliability through the program failure rate Λ_n.

Two general models linking bug exercise frequencies and occurrence rates have been formulated. In the *deterministic* model, each bug contained in the program is assumed to contribute to the program failure rate an amount *proportional* to its exercise frequency. The occurrence rate x of a bug is related to its exercise frequency f by

$$x = \gamma f, \tag{3.1}$$

where γ is a constant, equal for all bugs. This assumption is a strong simplification since the occurrence rate of a particular bug is not determined completely by its exercise frequency.

In the *stochastic* model the occurrence rate of bugs with given exercise frequency f are assumed to be drawn from a probability distribution whose *mean* is proportional to f. The deterministic model can be regarded as a limiting case of the stochastic model.

Whilst the stochastic model is more general than the deterministic, the additional parameters that are required to be calculated from the data raise the computational problems significantly. In this paper the simulated data experiment will be applied to the deterministic model.

3.3   Program failure rate

Consider the program at time τ 'now' after n failures. Assuming that T, the r.v. of the time between τ and the next failure, has the exponential distribution with parameter λ_n (Equation 2.3), then the program failure rate has the expected value,

$$\lambda_n = E(\Lambda_n). \tag{3.2}$$

The program failure rate after n failures ie equal to the sum of the occurrence rates of the bugs contained in the programme *at that time*. Hence

$$\Lambda_n \;=\; \text{program failure rate } \textit{at time } 0-$$

$$-\;\; \sum(\text{occurrence rates of the first } n \text{ bugs that have occurred } \textit{in time } [0,\tau]),$$

$$=\;\; \sum_{j=1}^{N} X_j - \sum_{i=1}^{n} \gamma f i \tag{3.3}$$

Here, $X_1, \cdots, X_n$ are i.i.d. (independent and identically distributed) random variables of the occurrence rates at time 0 of the N bugs initially in the program and the deterministic (DEF) model Equation (3.1) has been used to express the occurrence rates of the first n bugs in terms of their exercise frequencies.

Combining (3.2) and (3.3),

$$\lambda_n = \sum_{j=1}^{N} E(X_j) - \sum_{i=1}^{n} \gamma f i \qquad (3.4)$$

Using conditional probability arguments, expression (3.4) for the program failure rate λ_i after i failures can be cast into the form

$$\lambda_i = NF(a, b, \gamma) - \sum_{k=1}^{i} \gamma f_k \qquad (3.5)$$

where F is an integral function to be calculated numerically. Further details of the derivation of function F are given in Appendix B.

Expression (3.5) for the DEF model is to be compared with the corresponding expressions for the J-M model (Equation (2.6)), and the LSRG model (Equation (2.9)).

The DEF model expression (3.5) contains four parameters. N is the initial number of bugs in the program, γ is the parameter relating occurrence rates to exercise frequencies (Equation (3.1)), whilst a and b are the parameters of the parametric form for the probability density function pdf $(f|[0, \tau])$ that a bug has exercise frequency f given that it occurred in $[0, \tau]$.

Given that the first n bug failure times τ_i and their exercise frequencies f_i are known, then parameters N, γ, a, b, are estimated by the use of maximum likelihood functions $L1(a, b)$ and $L2(N, \gamma)$. The program failure rate λ_i can then be computed for any i from 1 to n, the end of the current data set. Details of the ML procedure are given in [3] and [4].

4　Simulated data experiment

To test the predictive capability of the DEF model against the J-M and LSRG models it is required that input data on failure times *and* bug exercise frequencies be available.

The exercise frequency model requires values of the exercise frequencies of the first n bugs. Data sets available in the literature do not contain program unit exercise frequencies and are therefore not adequate for testing purposes. To make further progress, it was necessary to obtain fresh data on failure times and exercise frequencies. Two methods are open. The first is to conduct a series of software running trials with appropriate software suitably instrumented to capture the necessary data. The second is to simulate the running of a software program and produce from the simulation the data on failure times and exercise frequencies.

Ideally the first method, capturing real software data, is the better solution. However the process of capturing failure and frequency data from industrial software is non-trivial (a project in itself). As an initial exercise it was decided to take the second option to test the model by simulating a running program and producing appropriate sets of failure time and exercise frequency data. This approach suffers from the weakness that the very process of setting up a simulation to produce exercise frequency data carries modelling assumptions that includes an occurrence rate/exercise frequency relationship. However the strength of this relationship is under control of the simulation, as is the variation (spread) of the occurrence rates for a given exercise frequency. One purpose of the simulation experiment is to explore the sensitivity of the occurrence rate/exercise frequency relation, that is, how do models only using failure times perform in comparison with the exercise frequency model? One key advantage of the simulation procedure over industrial data is that failure data of *all* bugs are known, and therefore the 'true' failure rate of any simulated program run is known at all times. With industrial data, only the past is known with certainty!

Imagine a software program running in a certain environment from a time 0. The number of bugs in the software at $t = 0$ is assumed to be N (unknown). The occurrence times of the first n bugs are $\tau_1, \tau_2, \cdots, \tau_n$.

The simulation should have the following properties:

- The failure process of the simulated system is equivalent to a homogeneous Poisson process with rate equal to the sum of the occurrence rates of the bugs contained in the program at that time,

- The tendency for bugs with high exercise frequencies to have high occurrence rates. Occurrence rates of bugs with given exercise frequency f form a probability distribution whose mean is proportional to f and whose coefficient of variation is independent of f.

4.1   Simulation procedure

Each simulated failure data set was generated as follows.

Step 1:   the simulation parameters are chosen:

- the number N of bugs contained in the program at time 0,

- the number n of bugs which have occurred up to the present time,

- parameter values $\imath, \rho$ of the probability distribution $\text{pdf}^{(s)}{}_{\text{init}}(f)$ chosen from which the exercise frequencies f_i of each of the N bugs are drawn,

- parameter values δ, μ of the probability distribution $\text{pdf}^{(s)}(x|f)$ chosen from which the occurrence rates x_i of the bug with exercise frequency f_i are drawn. The strength of the relationship between the occurrence rates and exercise frequencies is governed by the choice of parameter δ.

Step 2: the exercise frequencies f_i of each of the N bugs are generated by drawing random values from the probability distribution $\text{pdf}^{(s)}_{\text{init}}(f)$.

Step 3: for each bug with exercise frequency f generated in step 2, an occurrence rate x is drawn at random from the probability distribution $\text{pdf}^{(s)}(x|f)$.

Step 4: the occurrence times τ_i of each of the N bugs are generated. The occurrence time of the bug with occurrence rate x is drawn at random from the exponential distribution with parameter x (Equation (2.1)). The occurrence times are then ordered from 1 to n;

$$0 < \tau_1 < \tau_2 < \cdots < \tau_i < \cdots < \tau_N \leq \tau$$

Step 5: the present time τ is fixed at $(\tau_n + \tau_{tn+1})/2$.

At this point the true program failure rate (of the simulated program run) at any time is known, being the sum of the rates of the known bugs that have failed up to the present time. For a given set of parameters an appropriate graph of program failure rate against failure number can be constructed.

4.2 Parametric forms and values for the simulation distributions

Parametric forms for $\text{pdf}^{(s)}_{\text{init}}(f)$ and $\text{pdf}^{(s)}(x|f)$ were chosen to satisfy requirements for flexibility, for which two-parameter distributions were considered, coupled with:

- for $\text{pdf}^{(s)}_{\text{init}}(f)$, the possibility of high variability in the bug exercise frequencies,

- for $\text{pdf}^{(s)}(x|f)$, the mean being proportional to f and coefficient of variation independent of f.

In addition, there were other technical restrictions from using the NAG algorithms. Weibull distributions were finally selected for both distributions. Further details are provided in [4].

This choice allowed the strength of relationship between bug occurrence rates and exercise frequencies to be varied through the parameter δ. Any desired degree of variability in the bug exercise frequencies can be achieved through the choice of the parameter ι. The means of $\text{pdf}^{(s)}_{\text{init}}(f)$ and $\text{pdf}^{(s)}(x|f)$ are proportional to ρ and μ respectively.

Table 1. Initial simulated data sets

Data set	N	n
1	40	10
2	40	20
3	40	30
4	200	50
5	200	100
6	200	150
7	500	125
8	500	250
9	500	375

Key: N, the number of bugs initially in the program

           n, the number of bugs occurring to the present time

5   Results

5.1   First data set

Initially nine data sets were generated by selecting N, the initial number of bugs and n, the number of bugs occurring to the present time, in three groups of three data sets. The data sets are labelled from 1 to 9, shown in Table 1.

The values for N cover small, medium and large 'bug sets'. Within each N, values for n are for $N/4$, $N/2$ and $3N/4$ giving failure processes where at the time 'now' one quarter, one half or three quarters of the bugs actually have occurred.

The choice of simulation distribution parameters ρ, μ, δ and $\imath$ was made to achieve the following aims:

- high variability in the N bug exercise frequencies, for which $\imath$ was chosen at random from a uniform $(1/3, 1/2)$ distribution for each data set 1 to 9,

- exercise frequency of the average bug in a convenient range $(0.2, 0.6)$, for which ρ was fixed at 0.1 for all data sets,

- reasonably strong relationship between bug occurrence rates and exercise frequencies assumed, for which δ was chosen at random from a uniform $(4,8)$ distribution for each data set 1 to 9,

Table 2. Simulation exercise frequency and occurence rate distribution characteristics

Data Set	N	n	$\mathrm{pdf}^{(s)}{}_{\mathrm{init}}(f)$		$\mathrm{pdf}^{(s)}(x\|f)$	
			mean	variation	mean	variation
1	40	10	0.47	3.8	$0.19f$	0.19
2	40	20	0.50	4.0	$0.18f$	0.23
3	40	30	0.23	2.5	$0.18f$	0.27
4	200	50	0.31	3.0	$0.19f$	0.19
5	200	100	0.20	2.2	$0.19f$	0.17
6	200	150	0.57	4.3	$0.19f$	0.18
7	500	125	0.35	3.2	$0.18f$	0.26
8	500	250	0.23	2.4	$0.19f$	0.17
9	500	375	0.25	2.6	$0.19f$	0.17

- occurrence rate of the average bug to be proportional to (and less than) the exercise frequency with a coefficient in the range (0.18, 0.19) over the data sets.

In Table 2, the mean and variation of $\mathrm{pdf}^{(s)}{}_{\mathrm{init}}(f)$ and $\mathrm{pdf}^{(s)}(x|f)$ is set out for each of the initial data sets 1 to 9. The effect of the simulation procedure has been to create the failure data set for nine different 'programs' each of which exhibits different variability in exercise frequencies and in bug occurrrence rates.

The strength of association between the exercise frequencies and occurrence rates for the simulation runs is governed by the value of the coefficient of variation of $\mathrm{pdf}^{(s)}(x|f)$ shown in Table 2. The order of the data sets from strong association downwards is ([5,8,9],6,[1,4],2,7,3).

5.2 Estimates for the current program failure rate

For each of these data sets the J-M, LSRG and the deterministic exercise frequency model (DEF) were used to estimate the correponding *current failure rate* λ_n of the program, that is the failure rate immediately following the manifestation of n faults. Data available to the models consists of bug failure times and exercise frequencies, known from the simulation runs. This data is used by the models to estimate the current program failure rate. Model estimates may be compared with the 'true' failure rate computed from the simulation data which uses the *known remaining* $N-n$ bug occurrence rates rather than the past n failure times and execution frequencies. The simulation data, being complete, does give an advantage over 'real' failure data which is always incomplete.

Table 3. Error factors in predicting program failure rate λ_n from the true value

Data set	1	2	3	4	5	6	7	8	9
n	10	20	30	50	100	150	125	250	375
True λ_n	0.266	0.041	0.0116	1.82	0.347	0.00402	3.88	1.24	0.0874
J-M factor	1.2-	∞	∞	1.4+	1.3+	∞	1.0+	2.1-	∞
LSRG factor	1.2-	2.9+	1.6-	1.4+	1.8+	1.1+	1.2+	1.2-	1.0-
DEF factor	1.3+	4.0+	1.7+	1.9+	1.8+	1.1-	1.3+	1.1+	1.3-

Key: '+' means that the predicted rate is *higher* than the true rate by the given factor
'-' means that the predicted rate is *lower* than the true rate by the given factor

Two varieties of the DEF model were computed, depending upon the distributions taken (see Appendix B) and the results were essentially the same, so only a single DEF result is reported here. The J-M model was found to under-estimate the failure rate on six of the nine data sets, including four at which the failure rate prediction is zero (sets 2,3,6 and 9). This optimistic result is not surprising since there is high variability in the bug occurrence rates in the simulations. This feature was reported on in Section 2.3.

There was little to choose between the estimates given by the LSRG and the DEF models. The error factors by which the model predictions differ from true values are shown in Table 3. Error factors are defined by

$$\left|\frac{\max\left(\lambda_n(\text{true}), \lambda_n(\text{model})\right)}{\min\left(\lambda_n(\text{true}), \lambda_n(\text{model})\right)}\right|. \tag{5.1}$$

Any advantage lies with LSRG rather than the DEF model, perhaps surprisingly, suggesting a rather more robust model against data taken from random sources.

The LSRG model estimates the program failure rate using failure time data only. This suggests that the effect on model performance of knowing occurrence rates is limited. The limitation may be due either to the fact that neither LSRG or DEF models are good (relative errors are typically in the range 10-90% going up to 300%) or that the linkage between occurrence rates and exercise frequencies is too weak for the DEF model to exploit. The data in Table 3 does not show a clear improvement in DEF estimates where the strength of association is greater.

It should be noted that the estimates given in Table 3 are for the current failure rate following n failures. Predictions of *future* failure rate following future unknown failures are not given: the prime purpose of these experimental trials is to test the DEF model against the actual failure rate data using the LSRG model as a benchmark. Unless there are significant advantages in the exercise frequency model accuracy against LSRG, the additional data gathering and computational complexity would not be justified.

5.3 Retrodictive reliability growth tracking model performance

A second measure of model performance is to compare their ability to *track* the reliability growth curve, that is the successive program failure rate values λ_i from the *first* failure occurrence to the n'th (currrent) occurrence, for each simulation. Here, the model parameters of the various models are computed *once* using the failure data up to the n'th occurrence. Model expressions are now known, and their program failure rates computed as a function of failure number from 1 to n. The result can be shown in the form of a graph in which each model tracks the reliability *growth* of the program in time as failures occur and are removed. Calculations were performed for the nine initial data sets.

As before, estimates of λ_i from the J-M and LSRG models remain based on failure time data τ_1, τ_2, $\cdots$,τ_n. The DEF model estimates are based on the failure time data together with the exercise frequency data f_1, f_2, $\cdots$, f_n. Again the two DEF models produced essentially identical estimates.

The result for the smallest data set 1 is illustrated in Figure 6. It shows the tracking estimates of the models against the true reliability growth curve for $N = 40$, $n = 10$. For this particular set only, the LSRG and J-M models gave similar results, the DEF result lying between LSRG/J-M and True. The reliability growth tracking is *retrodictive* rather than *predictive*, in the sense that reliability estimates are limited to execution time up to the n'th failure (now), and do not predict reliability growth into the future. The test is to see whether the DEF model possesses any significant advantage in reliability growth tracking to offset the additional model and computational complexity.

The corresponding result for data set 2, $N = 40$, $n = 20$, is illustrated in Figure 7. For this case (and with all other data sets 3-9) the LSRG model differed from the J-M prediction. It is clear that the J-M model, by definition a straight line result, does not track the true growth curve closely. The DEF model tracks True closely, with the LSRG model homing in to it as the failure number increases.

The remaining data sets gave broadly similar graphs, with the LSRG and DEF models tracking the true reliability growth curve much more efficiently than J-M. For large N and n, it becomes difficult to distinguish

        A.W. Andrews and R.J. Cole

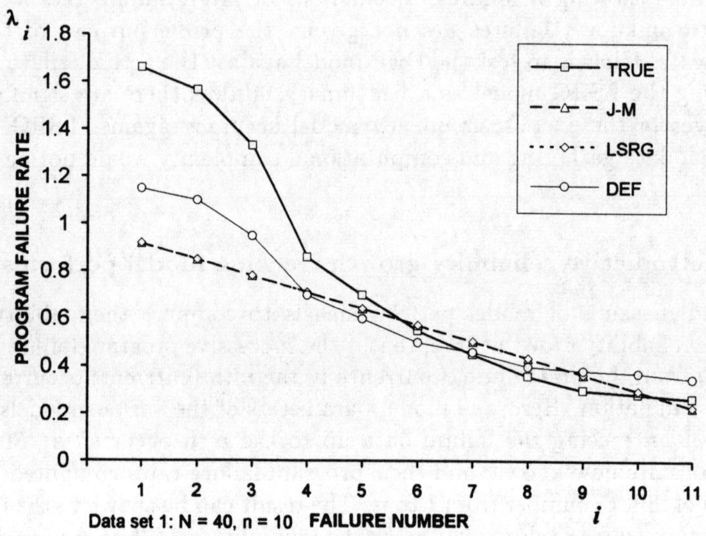

Figure 6. True λ_i and model estimates of λ_i against failure number

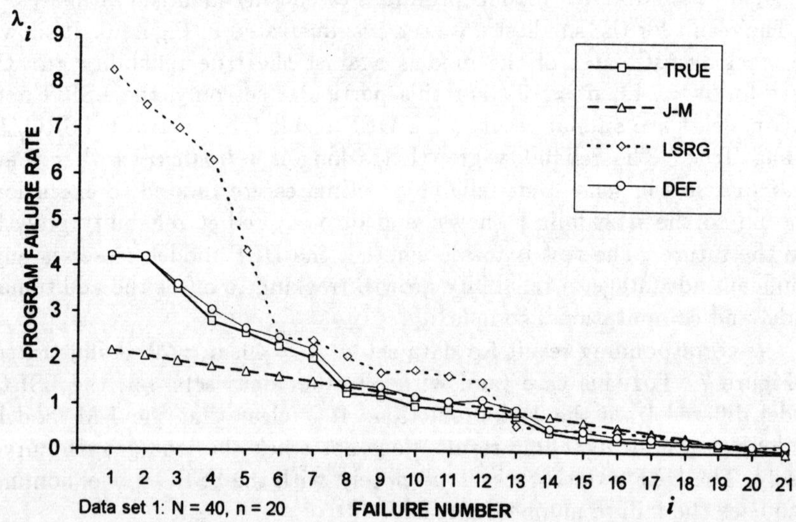

Figure 7. True λ_i and model estimates of λ_i against failure number

Table 4. Comparison of tracking performance of LSRG and DEF models: average of λ_i from $i = 1$ to n, for data sets 1-9

				Data	Set				
	1	2	3	4	5	6	7	8	9
LSRG	1.32	1.97	1.48	1.37	1.16	1.18	1.17	1.16	1.12
DEF	1.25	1.68	1.19	1.31	1.19	1.04	1.13	1.14	1.09

the model performance visually because the program failure rate, the slope of the graph, becomes very small as failures occur and are removed. Instead, the comparative performance of the LSRG and DEF predictions to the true value may be quantified for each data set by calculating the error factors for λ_i averaged over all i, for each data set separately, shown in Table 4.

The DEF model tracks the growth curves slightly more closely than LSRG, the improvement being marked in data sets 2,3 and 6, less significant in sets 1,4 and 7, and of little significance in sets 5,8 and 9.

5.4 Extended data set

The initial nine runs were expanded into twenty seven simulations to compare further the LSRG and DEF models for the current program failure rate λ_n. The overall results indicate that the predictions of the two models were again close with the LSRG slightly more accurate on average. For both models, predictions for the *largest class* data sets $[N = 500,\ n = (125, 250, 375)]$ were more accurate than for the *medium class* $[N = 200,\ n = (50, 100, 150)]$, and were more accurate for the *medium* than the *smallest class* $[N = 40,\ n = (10, 20, 30)]$.

6 Conclusions

In simulations containing high variability of bug occurrence rates, the Jelinski-Moranda model performed poorly, confirming that its assumption of equal bug occurrence rates leads to optimistic program reliabiliy estimates. The LSRG and DEF models produced similar results with the LSRG model having an edge in predicting current failure rate but the DEF model the edge in tracking the intermediate failure rates more closely.

It might be expected that the *stochastic* exercise frequency model (SEF) would be more likely to produce better predictions than the DEF, since it captures the *tendency* of bugs with high exercise frequencies to have high occurrence rates. However the search for the maximum likelihood function for the DEF takes place in two-parameter space whilst that for the SEF takes place in three-parameter space. This latter search leads to significant increased computational complexity with associated breakdown of the numerical algorithms. Further work is needed to overcome these numerical problems and also to investigate the extent to which program exercise frequencies can truly be reflected in bug occurrence rates.

Bibliography

1. Abdel-Ghaly, A.A., Chan, P.Y. and Littlewood, B. (1986). Evaluation of competing software reliability predictions. *IEEE Transactions on Software Engineering, SE-12*, 950-967.

2. Adams, E. (1984). Optimising preventative service of software products. *IBM Journal of Research and development, 1*, 1-14.

3. Andrew, A.W., Cole, R.J. and Gomatam, J. (1990). Modelling software reliability from run-time data. *Proc. Third European Conference on Mathematics in Industry*, pp. 225-231.

4. Andrew, A.W. (1990). A class of software reliability growth models based on exercise frequencies. PhD Thesis, Glasgow Caledonian University.

5. Bendell, A. and Mellor, P. (Editors). (1986). *State of the Art Report on Software Reliability*, Pergamon Infotech.

6. Jelinski, Z. and Moranda, P.B. (1972). Software reliability research. *Statistical Computer Performance Evaluation*, Editor: W. Freiberger, Academic Press, 455-484.

7. Littlewood, B. (1981). Stochastic reliability growth: A model for fault removal in computer programs and hardware designs. *IEEE Transactions on Reliability, R-30*, 313-320.

Appendix A Notation used

τ_1 execution time to first failure (from an origin 0)

τ_i execution time to ith failure (from the origin 0)

τ elapsed execution time 'now' (from the origin 0),

T r.v. time to next failure,

t realisation of r.v. T

ϕ	occurrence rate of a bug
Φ	r.v. occurrence rate of a bug
$\oslash$	(constant) occurrence rate of a bug
ϕ_i	occurrence rate of the i'th bug
N	initial number of faults in the program
n	number of failures up to time τ
λ	failure rate of the program
λ_n	failure rate of the program after n failures
Λ_n	r.v. failure rate of the program after n failures
$\mathrm{gamd}(x, \alpha)$	gamma pdf, $= x^{a-1} \exp(-x)/\Gamma(\alpha)$
$\mathrm{pdf}(\phi)$	$\beta\, \mathrm{gamd}(\beta\phi; \alpha)$: distribution of LSRG inital bug occurrence rates
f	exercise frequency of a bug
f_i	exercise frequency of the i'th bug
γ	constant of proportionality between exercise frequencies and occurrence rates

$\mathrm{pdf}^{(s)}(x\mid f)$ simulation distribution of occurrence rate x of bug with *given* exercise frequency f,

$$= \mathrm{Wiebull}\,(x; \delta, \mu f)$$

$$= \frac{\delta}{\mu f}\left(\frac{x}{\mu f}\right)^{\delta} - \exp\left[-\left(\frac{x}{\mu f}\right)^{\delta}\right], \delta, \mu > 0$$

$\mathrm{pdf}^{(s)}{}_{\mathrm{init}}(f)$ simulation distribution of exercise frequencies of each of the N bugs Pr(bug chosen at random from among the initial N present has exercise frequency f)

$$= \mathrm{Weibull}(f; \imath, \rho)$$

$$= \frac{1}{\rho}\left(\frac{f}{\rho}\right)^{\imath-1} \exp\left[-\left(\frac{f}{\rho}\right)^{\imath}\right], \imath, \rho > 0$$

$\text{pdf}_{\text{xinit}}(x)$ is the probability density that a bug chosen at random has occurrence rate x.

$\Pr\{[0,\tau]|f\}$ Prob(bug occurs given that it has exercise frequency f)

$\Pr\{[0,\tau]\}$ Prob(bug chosen at random occurs in $[0,\tau]$)

$\text{pdf}(f|[0,\tau])$ Prob(bug has exercise frequency f given that it has occurred in $[0,\tau]$)

$$= \text{log normal } (f;\xi,\sigma^2)$$

$$= \frac{1}{\sigma f \sqrt{(2\pi)}} \exp\left[-\tfrac{1}{2\sigma^2}(\ln f - \xi)^2\right], -\infty < \xi < \infty$$

OR $\text{pdf}(f|[0,\tau])$ $= \text{invgauss } (f;a,b)$

$$= \left(\frac{b}{2\pi f^3}\right)^{\frac{1}{2}} \exp\left[-\frac{b}{2a^2 f}(f-a)^2\right], a>0, b>0$$

a,b parameters of the DEF distribution $\text{pdf}(f|[0,\tau])$ for inverse Gauss

ξ,σ parameters of the DEF distribution $\text{pdf}(f|[0,\tau])$ for lognormal

Appendix B Derivation of failure rate for DEF model

We start with the failure rate expression (3.4), namely

$$\lambda_n = \sum_{j=1}^{N} E(X_j) - \sum_{i=1}^{n} \gamma f i. \tag{B1}$$

In (B1) and expressions for $E(X_j)$ is obtained as follows;

$$E(X_j) = \int_0^\infty \text{xpdf}_{\text{xinit}}(x)dx, \tag{B2}$$

where $\text{pdf}_{\text{xinit}}(x)$ is the probability density that a bug chosen at random has occurrence rate x. This can be related to the corresponding pdf for exercise frequency through the relation, using (3.1),

$$\mathrm{pdf}_{\mathrm{xinit}}(x) = \frac{1}{\gamma}\mathrm{pdfinit}\left(\frac{x}{\gamma}\right), \tag{B3}$$

giving

$$E(X_j) = \int_0^\infty \left(\frac{x}{\gamma}\right)\mathrm{pdf}_{\mathrm{init}}\left(\frac{x}{\gamma}\right)\mathrm{dx},$$

$$= \gamma \int_0^\infty \mathrm{pdf}_{\mathrm{init}}(f)\mathrm{df}. \tag{B4}$$

In turn the expression for $\mathrm{pdf}_{\mathrm{init}}(f)$ is obtained through the following conditional probability argument.

Pr (bug occurs in $[0,\tau]$ with exercise frequency f) =

= Pr(bug chosen at random from among the initial N present has exercise frequency f) *Pr(bug occurs given that it has exercise frequency f),

= Pr(bug chosen at random occurs in $[0,\tau]$) *Pr(bug has exercise frequency f given that it has occurred in $[0,\tau]$),

from which we obtain

$$\mathrm{pdf}_{\mathrm{init}}(f) = \frac{\mathrm{Pr}\{[0,\tau]\}\mathrm{pdf}(f|[0,\tau])}{\mathrm{Pr}\{[0,\tau]|f\}}. \tag{B5}$$

Now from (2.1)

$$\mathrm{Pr}\{[0,\tau]|f\} = 1 - \exp\left(-\gamma f\tau\right), \tag{B6}$$

yielding

$$\mathrm{pdf}_{\mathrm{init}}(f) \propto \frac{\mathrm{pdf}(f|[0,\tau])}{1 - \exp\left(-\gamma f\tau\right)}. \tag{B7}$$

At this stage the program failure rate (B1) is connected directly to expression (B7). The parametric form for $\mathrm{pdf}(f|[0,\tau])$ needs to be chosen to be flexible, mathematically tractable and result in a proper distribution over $[0,\infty]$. The forms finally selected with these desirable characteristics included the log normal and inverse Gaussian, each having two parameters. These are defined in Appendix A.

In either case the program failure rate parameter immediately after i failures is given using (B1), (B4) and (B7) by an expression of the form

$$\lambda_i = \mathrm{NF}(a, b, \gamma) - \sum_{k=1}^{i} \gamma f_k, \tag{B8}$$

where F is an integral that is computed numerically once parameters are known, a and b being parameters of the parametric form for $\mathrm{pdf}(f | [0, \tau])$ (a and b being replaced by ξ and σ in the case of the inverse Gaussian distribution).

Given that the values of the first n bug failure times τ_i and their exercise frequencies f_i are known, then maximum likelihood functions $L1(a, b)$ and $L2(N, \gamma)$ are used to estimate the parameters a, b, N, γ. Finally the program failure rate λ_i can be computed for any i from 1 to n, the end of the current data set.

Further analytic and computational details are to be found in [4].

Towards verified systems:
the SAFEMOS project

Jonathan P. Bowen and He Jifeng

*Oxford University Computing Laboratory, Programming Research Group,
Oxford*

Roger W.S. Hale and John M.J. Herbert

SRI International Cambridge Research Centre

Abstract The collaborative safemos project has investigated the formal development of embedded systems from specification through to a real-time programming language, compilation to object code and the formal design (and even automatic compilation) of a hardware machine to execute that code. The project has used Occam and the Transputer as an inspiration for its investigations, with real-time extensions where required. HOL has been used for mechanical verification where appropriate. A close liaison with the related collaborative European ESPRIT **ProCoS** project has been maintained to ensure that research on both projects is coordinated. This paper gives an overview of the work of the project with particular regard to the mathematical techniques used for the specification and verification process.

1   Background

In the last decade, the use of software in safety-critical systems has increased by around two orders of magnitude. However current widely used development techniques are still sadly lacking in their ability to avoid the occurrence of errors in such systems. Formal mathematically based methods provide one means to avoid the introduction of errors into systems at the design stage by increasing the preciseness of descriptions earlier on in the process and allowing the possibility of proven transformations and relationships between the specification and implementation of a system.

There is currently great interest in both academic and industrial circles in the issues involved in the use of formal methods but the techniques still need further investigation and promulgation to make their widespread use a reality [12, 13]. A number of safety-related standards, are currently being introduced and the recommendations in these could well have a significant impact on the use of formal methods in the development of safety-critical systems. Many standards in this area are now mentioning formal methods as one option to improve the correctness of safety-related software [5]. A few, such as the UK MoD 00-55 draft standard [41], even mandate the use of formal methods although there is still much debate and industrial resistance.

This provides the setting into which the research work briefly presented in this paper is intended to fit in years to come. Here we concentrate on an overview of mathematical techniques used by the **safemos** project intended to aid formal development of software and hardware for embedded high integrity systems.

2   Project overview

The collaborative UK IED (Information Engineering Directorate) **safemos** project (1989–1993) has investigated techniques to aid the formal verification of mixed hardware/software systems. Aspects of system specification and verification from an abstract formal description of the system down to the underlying hardware have been addressed, with particular regard to real-time issues.

The project was influenced and inspired by the simplicity of the Occam programming language [36], with its well established formal underpinning of the process algebra CSP (Communicating Sequential Processes) [31] and a wealth of algebraic laws [47] to aid formal transformation of programs. In addition, the Transputer microprocessor developed by Inmos [37] provides a platform for the implementation of Occam programs.

The HOL (Higher Order Logic) [20, 18, 52] theorem proving system was used to perform machine-checked proofs. The HOL system provides an LCF-style theorem proving environment [21, 45] and supports a version of classical higher-order logic based on Church's formulation of simple type theory [2, 16]. HOL includes ML as a metalanguage: the ML language was originally developed as part of LCF, but is now an independent programming language in its own right [40]. It is an eager-evaluation functional language with a polymorphic type discipline.

Some use of the formal notation Z has been made on the project. Z is based on Zermelo-Fraenkel set theory and first order predicate calculus [50], with the addition of the schema 'box' notation to aid the structuring of the large amount of detailed mathematics that is necessary to specify systems of a realistic size. An advantage of a formal notation like Z is that it has been accepted for international standardisation by ISO which is likely to lead to wider industrial acceptance [14]. Industrial practitioners are often reluctant to use raw mathematics and standardisation is a major contributor to aiding its infiltration in certain sectors such as those involved with safety-critical systems [12].

The work described here has been undertaken by the following industrial and academic partners in the UK:

- Inmos Limited, Bristol;

- SRI International Cambridge Computer Science Research Centre;

- Oxford University Computing Laboratory, Programming Research Group;

- University of Cambridge, Computer Laboratory.

The project has aimed to address some of the problems facing the designers and users of microprocessors and micro-controllers that are arising as the complexity and power of these devices increase. Microprocessors are being used to perform increasingly complex tasks; as a result the ability to ensure correct design by traditional design techniques, centered around experimental testing, is becoming problematic.

The use of formal design methods seems to offer a way out of this situation by providing a design methodology which prevents the introduction of errors into designs through the rigorous use of proof techniques to validate designs against specifications. The results of the **safemos** project aim to demonstrate the feasibility of these methods for real-time control systems. In the future, such real-time controllers will increasingly consist of processors running embedded programs along with specialised interface hardware. Thus the project has addressed the problem of verifying both hardware and software.

2.1 Goals

The original three goals of the project, started in 1989, were [48]:

1. to demonstrate that it is feasible and commercially advantageous to verify systems containing both hardware and software by machine checked formal proof;

2. to develop the methodology and tools needed for performing such verifications and for estimating their costs;

3. to gain improved scientific understanding of the practical use of existing formal methods and tools, including HOL [20], Z [51] and CSP [31].

These goals have been addressed by designing an Occam-like real-time language, a program verifier for that language, a verifiable Transputer-like processor design and a verified translator to compile the real-time language into the processor instruction set. A simple demonstrator example has been undertaken to show how a program can be verified to meet a specification.

A major research challenge was the co-ordination of the diverse formal proofs needed for diverse system verification within a coherent and uniform framework. The aim of the **safemos** project has thus been to develop formal methods for reasoning about Occam-like programs running on a Transputer-like processor, in the CSP tradition. While perfect compatibility between different components is very hard to achieve on a collaborative project at geographically separate sites, particularly one undertaking research, the early selection of an existing, and real, language and processor, together with the use of a single mechanical verification support tool, has prevented any site from becoming too devolved in their work. The following sections outline some of the most important work areas undertaken on the project.

Other related work is being carried out elsewhere, most notably, on the European collaborative ESPRIT Basic Research **ProCoS** project [8, 9], and at CLInc, a company in the US largely dedicated to the development and use of the Boyer-Moore theorem prover [17, 42]. Contact has been maintained with both these efforts; in particular, the **ProCoS** project is also studying Occam-like languages and Transputer-like machines, so work on both projects is directed towards a common goal. However on the **safemos** project there is a greater concentration on mechanically assisted proofs centred around a single tool, namely HOL, and a more balanced study of hardware as well as software verification issues.

2.2   The SAFEMOS tower

In designing a system, a number of different levels of abstraction must normally be considered. For example, a designer must transform a (possibly non-executable) *specification* into an *implementation* in the form of an executable program; a compiler must automatically convert a high-level program into low-level object code; and the underlying hardware must correctly execute that code using simple logic gates and latches (for example). The transformations should ideally be error-free and techniques to avoid the introduction of errors during this process are obviously desirable. The use of formal methods is one such technique since they reduce the ambiguity in the process, help to increase human understanding and allow formal proofs, reasoning and even calculation to be undertaken. Mechanical support for the process reduces the chance of human error which is highly likely in all but the simplest endeavours.

The **safemos** project has considered such transformations and mechanisation at a number of different levels:

- System specification: State-transition Assertions (STA), outlined in section 3, and Timed Transition Systems [28], described in [23].

- Compilation using an Occam-like sequential programming language and Transputer-like processor: Interval semantics based on Interval Temporal Logic (ITL) [22, 24, 43].

- Microprocessor design: Incremental framework using Higher-Order Logic (HOL) [20].

- Hardware compilation: A novel refinement algebra approach via a normal form as outlined in section 9, allowing the possibility of hardware/software co-design.

In an ideal world, different levels of abstraction would interface together to form a seamless 'tower' of design descriptions. Each level moves towards the final implementation by being assigned a semantics that is equivalent or 'better' than the previous level w.r.t. some refinement ordering. (Here, 'better' means being more deterministic and/or terminating more often.) This ideal has yet to be fully achieved in practice although the work of the related **ProCoS** project continues to strive for this goal.

A uniform proof environment, namely HOL (Higher-Order Logic), has been used for the mechanisation of proofs on the project, and the adoption of a common tool helps in the linking of different levels of abstraction. This has been the experience of the CLInc effort [17], as previous mentioned, who use the Boyer-Moore theorem prover for proofs of a number of related levels in both software and hardware.

The rest of the paper provides details of specific aspects of the work of the project.

3 State transition assertions

A graphical state-transition approach to specifying hard real-time reactive systems has been developed [19]. This is refined to a formal notation based on sentences called *State Transition Assertions* (STAs). These have a set-theoretic semantics that can be used to justify various laws, which combine aspects of *Interval Temporal Logic* (ITL) [22, 24, 43] and and Hoare Logic [30]. The semantics of programs can also be represented by sets of STAs, and then verification is performed by using laws to combine the STAs from the program to obtain the specification.

3.1 Hard real-time

Hard real-time systems are required to meet explicit timing constraints, such as responding to an input within 100 milliseconds of a change. The temporal requirements are an essential part of the required behaviour, not just a desirable property as in the case of *soft real-time* systems. The work aims to combine hardware and software verification techniques to produce a formal method for hard real-time programming. In this method, programs are refined from specifications consisting of a kind of state transition diagram. These diagrams have a precise semantics and the verification that programs implement them is by machine checked formal proof.

A key element of the method is that the program semantics used for verification is determined by what happens when the compiled program runs on the processor being used. This is achieved by defining the semantics via the compiler and processor specification.

3.2   Method

This work provides a possible foundation of a formal method for developing hard real-time programs. In summary, the method is as follows:

1. Write the specification using annotated state-transition diagrams and interpret them as sets of state transition assertions (STAs).

2. Develop a program by identifying nodes in the diagram with sets of processor states.

3. Verify the program by showing that its transitions (which are mechanically derived from the compiler and processor specification) entail the required transitions using laws for combining STAs.

Specifications are formalised as predicates on machines. A typical specification involves a number of transitions between wait states. These can be represented using a *state transition assertion* (STA), that combines aspects of the 'leads-to' and 'until' operators of temporal logic. The general form of an STA is:

$$M \models A \xrightarrow[P]{Q} B$$

where:

- $M : inputs \longrightarrow state \longrightarrow state$ is a machine;

- $A : state \longrightarrow bool$ is called the *state precondition*;

- $B : state \longrightarrow bool$ is called the *state postcondition*;

- $P : \text{seq } inputs \longrightarrow bool$ is called the *input precondition*;

- $Q : \text{seq } state \longrightarrow bool$ is called the *output postcondition*.

(*bool* represents Boolean values.)

The intuition behind state transition assertions is as follows: if $\mathcal{M}$ is in a state satisfying A and a sequence of inputs arrives that satisfies P, then a state satisfying B will be reached and the sequence of intermediate states will satisfy Q.

A simple example of a multiplier program has been used to demonstrate the technique, mechanically checked using HOL, and is presented in [7].

4 Real-time programming language

The **safemos** programming language, SAFE, is a real-time sequential imperative language with input and output constructs and with deadline constraints. The syntax of SAFE processes is given by the following BNF, where p is a process, e an expression, b a Boolean, x a variable and t a time.

$$p \quad ::= \quad \text{SKIP} \mid \text{STOP} \mid x :=_t e \mid \text{READ}_t\, x1\, x2 \mid \text{WRITE}_t\, x\, e \mid$$
$$p1\,;\, p2 \mid \text{IF}_{t1,t2}\, b\, p1\, p2 \mid \text{WHILE}_{t1,t2}\, b\, p \mid \text{LOCAL}\, x\, p$$

This syntax has been formally represented in HOL as a recursive type definition.

The process SKIP does nothing and terminates immediately. STOP holds the execution forever. The assignment, $x :=_t e$, assigns the expression e to the variable x, and takes at most t time units to complete. Input and output are also time-bounded. The input process, $\text{READ}_t\, x1\, x2$, assigns the current input on port $x1$ to variable $x2$, and the output process, $\text{WRITE}_t\, x\, e$, outputs the expression e on port x. Input and output in SAFE are memory-mapped; that is, the input and output constructs are simply assignments to I/O registers and do not perform synchronisation. Synchronisation may be achieved by means of protocols implemented in SAFE [15].

Sequential composition, $p1\,;\, p2$, does not introduce any time overhead. For the conditional, $\text{IF}_{t1,t2}\, b\, p1\, p2$, no more than $t1$ time units may be used in testing the Boolean expression b, and no more than $t2$ units following the termination of either $p1$ or $p2$. The loop, $\text{WHILE}_{t1,t2}\, b\, p$, is a conventional while loop in which at most $t1$ time units may be used for testing the condition b, and at most $t2$ units for reinitiating the loop. Finally, the construct LOCAL x p makes the variable x local to process p. A very similar real-time language has also been developed on the **ProCoS** project [25].

We assume here that time constraints are measured in machine cycles but, in general, any appropriate time units could be used – even real time, given a suitable mapping between states and real times.[1] Whether or not a particular time constraint can be achieved in reality obviously depends on the complexity of the expression and on the underlying architecture. The

---

[1] We assume that a program necessarily executes in a sequence of discrete steps.

constraints are therefore checked by the compiler; values that are too small are rejected.

5   Interval semantics

We have defined the formal semantics of SAFE in HOL. It is a model-based semantics based on ITL, which captures timing properties in a natural way and also permits a uniform treatment of programming and assembly language constructs. This facilitates compiler verification.

Each SAFE process is identified with a predicate on intervals, as described below. An interval is a non-empty sequence of states and may be either finite or infinite; its length represents the number of time steps between start and finish.

In the HOL definitions below σ stands for a (finite or infinite) interval; $\text{len}\ \sigma$ for its length; $\text{init}\ \sigma$ and $\text{last}\ \sigma$ for the initial and final states of σ; and $\text{pfx}\ i\ \sigma$ and $\text{sfx}\ i\ \sigma$ for the ith prefix and ith suffix subinterval, respectively. Let us also write $\mathcal{M}^{\text{p}}\ p$ for the meaning of a process p, and $\mathcal{M}^{\text{e}}\ e$ and $\mathcal{M}^{\text{b}}\ b$ to denote the meanings of expression e and Boolean b, respectively.

The process SKIP is identified with the predicate Skip, which is true on any zero-length interval.

$$\mathcal{M}^{\text{p}}\ \text{SKIP}\ \sigma \stackrel{def}{=} \text{Skip}\ \sigma$$
where
$$\text{Skip}\ \sigma \stackrel{def}{=} (\text{len}\ (\sigma) = 0)$$

The process STOP never terminates and keeps the internal state, i.e. the set V of program variables and output ports, stable (constant).

$$\mathcal{M}^{\text{p}}\ \text{STOP}\ \sigma \stackrel{def}{=} \text{len}\ (\sigma) = \infty \land \text{Stb}\ \text{V}\ \sigma$$
where
$$\text{Stb}\ E\ \sigma \stackrel{def}{=} \forall e \in E, \forall n \bullet \sigma_n(e) = \sigma_0(e)$$

The assignment process must satisfy the appropriate timing constraint, in addition to assigning the value of an expression to a variable. It also carries the unspoken assumption that the part of the internal state not explicitly changed remains constant.

$$\mathcal{M}^{\text{p}}\ (x :=_t e)\ \sigma \stackrel{def}{=} (x :=_{0,t} (\mathcal{M}^{\text{e}}\ e))\sigma$$
where
$$x :=_{t1,t2} e\ \sigma \stackrel{def}{=} (\text{last}\ \sigma)x = e(\text{init}\ \sigma)\ \land$$
$$(t1 \leq \text{len}\ \sigma \leq t2)\ \land$$
$$(\text{Stb}\ (\text{V} - \{x\})\ \sigma)$$

Input and output are special forms of assignment.

Assignment is easily generalised to sequences of variables and expressions, and a delay, then, is just an assignment with null arguments:

$$\texttt{Dly}_{t1,t2} \stackrel{def}{=} \langle\rangle :=_{t1,t2} \langle\rangle$$

The sequential composition of two processes is true on an interval if it can be chopped into two subintervals, such that the first process holds on the prefix and the second on the corresponding suffix. Alternatively, the first process may be non-terminating.

$$\mathcal{M}^{\mathrm{P}}\,(p1\,;\,p2)\,\sigma \stackrel{def}{=} (\mathcal{M}^{\mathrm{P}}\,p1\,;\,\mathcal{M}^{\mathrm{P}}\,p2)\sigma$$
where
$$(p1\,;\,p2)(\sigma) \stackrel{def}{=} p1(\sigma) \wedge \texttt{len}\,(\sigma) = \infty \ \vee$$
$$\exists\,i \bullet p1(\texttt{pfx}\ i\ \sigma) \wedge p2(\texttt{sfx}\ i\ \sigma)$$

The time-bounded conditional is defined in terms of a simple conditional with appropriate delays inserted.

$$\mathcal{M}^{\mathrm{P}}\,(\texttt{IF}_{t1,t2}\ b\ p1\ p2)\,\sigma \stackrel{def}{=}$$
$$((\texttt{Dly}_{0,t1}\,;p1\,;\texttt{Dly}_{0,t2}) \vartriangleleft b \vartriangleright (\texttt{Dly}_{0,t1}\,;p2\,;\texttt{Dly}_{0,t2}))\sigma$$
where
$$(p1 \vartriangleleft b \vartriangleright p2)\sigma \stackrel{def}{=} (b(\texttt{init}\ \sigma) \Rightarrow p1\ \sigma) \wedge (\neg b(\texttt{init}\ \sigma) \Rightarrow p2\ \sigma)$$

A simple loop predicate may be defined in the conventional way as the least fixed point of the function $\lambda\,X \bullet (p\,;X) \vartriangleleft b \vartriangleright \texttt{Skip}$.

$$\texttt{Loop}\ b\ p \stackrel{def}{=} \texttt{FIX}\,(\lambda\,X \bullet (p\,;X) \vartriangleleft b \vartriangleright \texttt{Skip})$$

where $\texttt{FIX}\ F$ denotes the least fixed point of a function F under the refinement ordering defined by $\sqsupseteq$, where

$$p1 \sqsupseteq p2 \stackrel{def}{=} \forall\,\sigma \bullet p1\ \sigma \Rightarrow p2\ \sigma$$

Using the choice operator, the least fixed point may be defined non-constructively in HOL and, when F is continuous, the fixed point may be proved equivalent to a limit of an approximating chain in the standard way. Since all SAFE constructs are continuous, the loop is indeed equal to the limit of iteration, as desired, and this has been verified using HOL.

The time-bounded while loop is constructed from the simple loop and appropriate delays.

$$\mathcal{M}^{\mathrm{P}}\,(\texttt{WHILE}_{t1,t2}\ b\ p)\,\sigma \stackrel{def}{=} (\texttt{Loop}\ b\ (\texttt{Dly}_{1,t1}\,;p\,;\texttt{Dly}_{0,t2})\,;\texttt{Dly}_{1,t1})\sigma$$

Note that the evaluation delay is of minimum length 1. This constraint avoids the possibility of having a non-terminating zero-length loop and is similar to the so-called 'non-Zeno' condition.

5.1   Algebraic laws

A library of algebraic laws which are generally useful in proofs concerning
the programming language has been formulated and proved correct. Below
are a few examples of such laws.

The empty interval Skip is 'better' (w.r.t. the refinement ordering relation $\sqsubseteq$) than an interval during which an expression E must remain stable:

$$\mathtt{Skip} \sqsupseteq \mathtt{Stb}\,E$$

The conjunction of two stability constraints is equivalent to a single constraint on the union of the arguments:

$$\mathtt{Stb}\,(E1 \cup E2) = \mathtt{Stb}\,E1 \wedge \mathtt{Stb}\,E2$$

The conjunction of a stable interval distributes through the sequential composition of two programs:

$$(p1\,; p2) \wedge \mathtt{Stb}\,E = (p1 \wedge \mathtt{Stb}\,E)\,; (p2 \wedge \mathtt{Stb}\,E)$$

An empty interval sequentially composed before or after a program is in-
distinguishable from the original program:

$$p\,; \mathtt{Skip} = p = \mathtt{Skip}\,; p$$

Sequential composition is associative:

$$p1\,; (p2\,; p3) = (p1\,; p2)\,; p3$$

All of these and many other laws have been formally derived in HOL from
the interval semantics described above. See [7] for further details.

The law (actually a "law schema") for composing assignments is more
complex than might initially be expected because of the need to treat both
I/O and internal variables in a consistent way. However, a function has
been implemented in HOL to take care of the complexity and automatically
compose assignments using this law. The function is perfectly rigorous; it
formally proves the required result. For example, given the composition
$x :=_{t1,t2} m\,; y :=_{t3,t4} x + n$, the function will prove the theorem (assuming
$x \neq y$):

$$x :=_{t1,t2} m\,; y :=_{t3,t4} x + n \;\sqsupseteq\; x,\, y :=_{t1+t3,t2+t4} m,\, m + n$$

The function may fail if the final values of the variables on the left of the
assignment are indeterminate. If r is an input port, it will not reduce the
composition $x :=_{t1,t2} e\,; y :=_{t3,t4} r$ without further information about the
input value.

These laws allow proofs about the compiler specification to be conducted at a higher level than would otherwise be the case by remaining in
the framework of the programming language itself where possible. They
could also be used for other purposes (e.g. program transformation for optimisation).

6   Assembly language

The machine has three registers, A, B, and C, a program pointer, P, and an addressable memory. An instruction *ins* has the effect of an assignment in which the machine state becomes some function of the old state.

$$state :=_{T(ins)} f(state)$$

Thus, the meaning $\mathcal{M}^i$ *ins* of an instruction *ins* may be defined in terms of the interval semantics above.

For example, the LDC instruction pushes a constant w onto the three-register stack:

$$\mathcal{M}^i(\text{LDC }w) \stackrel{def}{=} \text{A, B, C, P} :=_{T(\text{LDC})} w, \text{A, B, P}+1$$

The JMP instruction performs a jump by updating the program counter relation to the location after the instruction:

$$\mathcal{M}^i(\text{JMP }w) \stackrel{def}{=} \text{A, B, C, P} :=_{T(\text{JMP})} \text{A, B, C, P}+w+1$$

ADD (for example) is a two-operand instruction that pops the two input values off the register stack (i.e. from registers A and B) and pushes the result back onto the stack (i.e., into register A):

$$\mathcal{M}^i\text{ ADD} \stackrel{def}{=} \text{A, B, P} :=_{T(\text{ADD})} \text{A}+\text{B, C, P}+1$$

All the instructions for the machine have been specified in this manner [7]. Note that we ignore the variable length of optimised Transputer instructions for simplicity, but this has been handled elsewhere [3, 11, 34].

The behaviour of an assembly language program stored between locations $n1$ and $n2$ in ROM is just the combined effect of running the sequence of instructions between these locations:

$$\text{Run ROM }n1\,n2 \stackrel{def}{=} \text{Loop }(n1 \leq \text{P} < n2)\,(\mathcal{M}^i(\text{ROM}(\text{P})))$$

6.1   Z semantics

A machine semantics for a subset of the Transputer instruction set at the bit-level has also been specified using the formal Z notation [3, 51]. Z is more readable that HOL since it is designed for this purpose, so a Z specification was produced as an experiment in producing a formal description that is readable enough to be used as documentation for a suitably trained engineer.   For example, the state may be recorded in a *schema* box as follows:

$$
\begin{array}{|l}
\hline
\_State\_\_\_\_\_\_\_\_\_\_\_\_\_\_\_\_\_\_\_\_ \\
\mathbf{A},\mathbf{B},\mathbf{C} : Value \\
\mathbf{P} : Address \\
\mathbf{ROM},\mathbf{RAM} : Address \nrightarrow Value \\
clock : \mathbb{N} \\
\hline
\mathrm{dom}\,\mathbf{ROM} \cap \mathrm{dom}\,\mathbf{RAM} = \emptyset \\
\hline
\end{array}
$$

The address space of the program memory (**ROM**) and data memory (**RAM**) at modelled as partial functions and their domains must not overlap. The number of *clock* cycles since initialisation is also recorded as a natural number. During a change of state the program in ROM remains the same and an instruction always takes a non-zero number of clock cycles to execute:

$$
\begin{array}{|l}
\hline
\_\Delta State\_\_\_\_\_\_\_\_\_\_\_\_\_\_\_\_ \\
State \\
State' \\
cycles : \mathbb{N}_1 \\
\hline
\mathbf{ROM}' = \mathbf{ROM} \\
clock' = clock + cycles \\
\hline
\end{array}
$$

Schemas may be 'included' within other schemas and the after state is indicated by dashed variables by convention. The **LDC** instruction may then be specified as follows:

$$
\begin{array}{|l}
\hline
\_LDC\_\_\_\_\_\_\_\_\_\_\_\_\_\_\_\_\_\_\_\_ \\
\Delta State \\
w : Value \\
\hline
(\mathbf{LDC}, w) = M(\mathbf{ROM}(\mathbf{P})) \\
(\mathbf{A}', \mathbf{B}', \mathbf{C}', \mathbf{P}') = (w, \mathbf{A}, \mathbf{B}, \mathbf{P}{+}1) \\
\mathbf{RAM}' = \mathbf{RAM} \\
cycles = T(\mathbf{LDC}) \\
\hline
\end{array}
$$

M and T are functions that decode a binary opcode and return the timing for an instruction respectively. Note that LDC does not update the RAM contents.

Such a specification could be of use to a microprocessor designer, acting as the specification of the processor to be implemented. More recently we have investigated the embedding of Z within HOL which enables the efficient mechanisation of proofs about Z specifications [10].

7 Compiler specification

A compiler can be specified as a relation defined recursively over the syntax of commands and expressions.:

$$\mathcal{C}^{\mathrm{P}} \; q \; S \; n1 \; n2 \; \mathtt{ROM}$$

The relation is true if the instruction store $\mathtt{ROM}$ contains the compiled code for the process q starting at location $n1$ up to (but not including) location $n2$ under the symbol table S. We define the constraints on this relation for each high-level program construct. For example, $\mathtt{SKIP}$ may be implemented by simply making the finish address $n2$ of the matching object code the same as the start address $n1$:

$$\mathcal{C}^{\mathrm{P}} \; \mathtt{SKIP} \; S \; n1 \; n2 \; \mathtt{ROM} \; \stackrel{def}{=} \; n2 = n1$$

Sequential composition is implemented by placing the respective object code segments for the two compiled sub-programs contiguously together in the $\mathtt{ROM}$:

$$\mathcal{C}^{\mathrm{P}} \, (p1 \, ; \, p2) \, S \; n1 \; n2 \; \mathtt{ROM} \; \stackrel{def}{=}$$
$$\exists \, n \bullet \mathcal{C}^{\mathrm{P}} \; p1 \, S \; n1 \, (n1 + n) \; \mathtt{ROM} \; \wedge \; \mathcal{C}^{\mathrm{P}} \; p2 \, S \, (n1 + n) \, n2 \; \mathtt{ROM}$$

n is the offset from the beginning of the code of the location at which the two pieces of object code abut and is existentially quantified so that it is not directly visible in the combined object code.

Similar constraints have been formulated for all the program constructs and the compilation relation may be 'executed' (by theorem proving) in HOL to obtain the compiled code. Details are to be found in [7]. For those instructions that have time constraints, the total time required for the relevant sequence of instructions is checked. If this exceeds the bound, the compilation relation is false and the code fails to compile.

7.1 Correctness of compilation

The compilation of a process is correct if the behaviour of the compiled code is an implementation of the program behaviour. If the code for q is compiled in $\mathtt{ROM}$ between $n1$ and $n2$ using a symbol table S, then the execution of the instruction between $n1$ and $n2$ implements the behaviour of q under the data representation $\Theta \, S$:

$$\forall S \; n1 \; n2 \; \mathtt{ROM} \bullet \mathcal{C}^{\mathrm{P}} \, p \, S \; n1 \; n2 \; \mathtt{ROM} \; \Rightarrow$$
$$(\mathrm{P} :=_{0,0} n1 \, ; \, \mathrm{Run} \; \mathtt{ROM} \; n1 \; n2 \; \sqsupseteq \; (\mathcal{M}^{\mathrm{P}} \, p) \circ (\Theta \, S) \, ; \, \mathrm{P} :=_{0,0} n2)$$

The compilation scheme specified for the SAFE language has been mechanically proved correct w.r.t. this correctness criterion using algebraic

laws as previously described within the HOL theorem prover. The approach
for the verification draws on the ideas of Hoare [32]. A similar approach
has been adopted by the **ProCoS** project, although there the proofs have
largely been undertaken by hand only [34].

Here we have presented a simple unoptimised compilation scheme which
is desirable in a high integrity system to avoid errors. However optimisation is important in general and more efficient compilation strategies may
be attempted in a similar setting by adding further allowed compilation
relations (even including extra program constructs and machine instructions) without necessarily invalidating those that have already been proved
correct [26].

7.2 Rapid prototype compilation

It is relatively straightforward to produce a compiler which matches a compilation scheme such as that presented here very directly using a logic programming language like Prolog [4]. The formal compilation description for
each programming construct may be implemented as a Horn Clause. For
example, the clauses for `SKIP` and sequential composition may be implemented as follows in Prolog:

```
cp(skip,S,N1,N2,ROM) :- N2=N1.


cp(P1;P2,S,N1,N2,ROM) :-
    cp(P1,S,N1,N1plusN,ROM), cp(P2,S,N1plusN,N2,ROM).
```

What is more, it is even possible to produce a *decompiler* which takes the
object code (and the symbol table if available) and produces a matching
high-level program for non-optimised code [6]. This could be useful in
the verification and checking of safety-critical code where optimisation is
normally avoided anyway. A logic program specifies a relation in general
so such a program may also be used as a compiler *checker* taking a given
program, symbol table and matching object code and checking that they
are compatible with a formally specified and proven compiling relation.
Tools to help verify termination and the validity of the omission of the
occurs-check for Prolog implementations for such prototype compilers have
been produced elsewhere [38].

8   Microprocessor design

A machine code program must be executed on some underlying hardware,
normally based around a microprocessor in embedded real-time systems.
Important considerations for the microprocessor design are that it should

be predictable and extensible, and the mathematical models used must support these aims.

Predictable performance is important for hard real-time applications and this goal is met by using a synchronous memory, and by using models incorporating explicit time for the basic components. This means that the performance of the processor can be calculated in units corresponding to the period of the underlying synchronous clock. For example, a loadable register is defined in HOL as:

```
LOAD_REG(in:time->*, load, out)   def
                                  =
     (∀ t. out (t+1) = (load t => in t | out t))
```

The explicit timing information at the lowest level is then propagated up through the higher levels yielding hard real-time information at these levels.

8.1 Support for incremental design

The goal of an extensible design is supported by the incremental framework developed and used in the **safemos** project [29]. This builds on the idea in [53] for a general framework supporting the formal specification and verification of a range of processors. To support formal methods during the design process the models and techniques allow partial specification, extensions to a design and incremental verification of a design. The main elements of the framework are:

- uniform hierarchy of computation models;

- use of generic arguments;

- verification template;

- the use of incremental models and techniques.

The hierarchy of computation models follows the idea of a hierarchy of interpreters in [1], where an instruction at each level is interpreted by executing a series of instructions at a lower level, and these instructions are in turn executed by an interpreter at the next lower level. Generic arguments represent certain aspects of a design which can remain unspecified and be parameters of the design. For example, rather than specifying a 16-bit processor, the word size may be supplied as a parameter of the design; then a generic n-bit processor may be verified instead.

Having a general hierarchical model of computation such as an interpreter means that templates for specifying the instructions and interpreter can be provided; these encapsulate in a general theorem the combination of individual instruction correctness results at one level to derive correctness of the interpreter at the next higher level. To support flexibility in the specification and verification of the design, incremental models have been introduced replacing the previous functional interpreter models.

8.2 Relational interpreter framework

The basic model of system behaviour is a transition system where the next state is derived from the current state and environment (inputs). In a *relational interpreter* the effect of a certain transition can be defined as a predicate on the next state and the present state and environment. The behaviour of a transition system can be specified by the desired properties given as a set of (tag, predicate) pairs, and a selection function that indicates whether a particular property tag is chosen by a state and environment. The transition system for a certain set of properties must ensure that whenever the state and environment select a transition in the set then all predicates for the indicated transition must hold on the next state and the current state and environment. The definition in HOL is:[2]

```
TRANSITION_SYS(is_selected,prop_list)s e  def
                                           =
  (∀t.
    let a_tag = εtag. is_selected(tag,s t,e t)
    in
     (∀prop.
        prop MEM prop_list ∧ (FST prop = a_tag) ⟹
        SND prop(s(t + 1),s t,e t)))
```

The definition of `TRANSITION_SYS` gives an incremental model since more items can be incrementally added to the property list parameter as the specification is extended with more behaviour.

In addition to the specification of the microprocessor as relational interpreter using `TRANSITION_SYS`, an incremental model for the machine microcode has been devised. This allows the derivation of the following result:

```
ALL_UNIQUE (APPEND mcode1 mcode2) ⟹
  ROM(addr,out)(APPEND mcode1 mcode2) ⟹
    ROM(addr,out)mcode1
```

---

[2] Note that a `MEM` list is true if a is a member of list `list`; `FST` and `SND` access the first and second elements of a tuple. `is_selected(tag,s t,e t)` is true at time t, if `tag` is selected in state `s t` and environment `e t`. ε is Hilbert's choice operator in HOL. If a unique tag is chosen then the term εtag. `is_selected(tag,st,env)` (i.e. choose the tag such that ...) evaluates to this tag.

This states that subject to the condition `ALL_UNIQUE` which demands that no entries clash, a microcode ROM extended with the microcode set `mcode2` allows only behaviour which is compatible with the original set `mcode1`.

8.3 Verification

The relational interpreter framework and incremental implementation models have been developed in the context of doing a simple processor design [29]. As described in section 6, the state consists of a tuple of program counter P, three-element stack (A,B,C) and memory ROM, and there is an external signals environment represented by a variable **env**. For example, the semantics for the LDC instruction presented in section 6 can be described in HOL as follows:

```
LDC_SEM gen_rep ((P',A',B',C',ROM'),(P,A,B,C,ROM),env) =
   (let instr = ROM((ADDR_FN gen_rep)P)
     in
    let w = (ARG_FN rep)instr
     in
    ((P' = (ADD1 gen_rep)P) /\
     (A' = w) /\ (B' = A) /\ (C' = B) /\ (ROM' = ROM)))
```

where **gen_rep** contains the generic parameters of the design.

A simple microcoded machine has been designed to implement a complete instruction set [29], and has been shown to correctly implement a relational interpreter for the desired instructions.[3] The HOL description takes the following form:

```
(MICRO_MC MLIST_A gen_rep) ==>
    TRANSITION_SYS
    (IS_SELECT gen_rep,
    [LDC_op, LDC_SEM gen_rep;
     JMP_op, JMP_SEM gen_rep;
     ADD_op, ADD_SEM gen_rep;
     ... ])
    ((SIG_TUP5(PC,A,B,C,mem)) o
        (Temp_Abs(λ t. mp_index t = 0)))
    (e o (Temp_Abs(λ t. mp_index t = 0)))
```

---

[3]This result is subject to the synchronisation condition as described in [29].

The microcoded machine is described by `MICRO_MC` and is parameterised by the microcode `MLIST_A`. This means that the incremental models of the implementation and abstract machine can be used in tandem. The machine specification can be extended by adding more instructions to the property list argument of `TRANSITION_SYS`, and the microcode can be extended to implement these instructions thus extending the microcode parameter of `MICRO_MC`. The use of incremental models means that just these extensions need be verified, the previous results being inherited, and the new system can be verified in an efficient manner.

8.4   Inmos processor

In addition to the simple processor for the restricted machine instruction set, a more realistic processor for an enlarged instruction set has been designed and verified using HOL by David Shepherd at Inmos [7]. This was an important part of the project to demonstrate the usefulness of the formal methods developed by the more academic partners in an industrial setting. Inmos have been enthusiastic protagonists of formal methods: in the past they have won a UK Queen's Award for Technological Achievement jointly with Oxford University for their work on formally developing the microcode for the floating point unit of the T800 Transputer [49] and more recently they have applied formal techniques to critical parts of the T9000 Transputer such as the pipeline architecture and the associated virtual channel processor [39, 46]. The Inmos design produced on the **safemos** project is intended to be realistic in performance and complexity, as well as predictable and extensible.

The mathematical models underlying the Inmos processor are based on those described for the simple processor presented here, and thus support a generic, incremental approach to design specification and verification. In addition, there is an emphasis on automated proof using HOL which has resulted in the architectural approach employing generic parameters and a microcode verification tool as described in [7].

9   Normal form and hardware compilation

An alternative and novel approach to the implementation of a program on a microprocessor is to compile the program directly into hardware [44]. This technique was not foreseen at the start of the project, and as such is more speculative than the work presented in previous sections, but is presented here because we believe this will be an important development for the future. It has been made possible in a practical sense for small programs (such as those found in embedded systems) by the fast developing and exciting technology of Field Programmable Gate Arrays (FPGAs) which allows the configuration of a hardware circuit to be defined by binary code in a memory in a similar manner to more conventional object code.

The project has investigated an approach to such compilation via a *normal form* [35] which uses a very restricted subset of the high-level programming language being compiled. This splits the compilation process in two which both simplified the individual transformations between the levels and allows the possibility of compiling the normal form to either hardware or software or even a combination of the two.

A normal form program comprises three sequential programs where the first one designates the initial control state of the circuit normally taken from the environment (an assumption), and the last one the final state normally returned back to the environment (an assertion). The middle one is a loop with a multiple assignment as its body which specifies state change of the circuit during execution.

$$
\begin{aligned}
\mathcal{N}(n1,\, n2,\, K,\, C,\, V) \;&\overset{def}{=}\\
&\texttt{LOCAL}\ c\ (c :=_0 n1\,;\\
&\texttt{WHILE}\,(c \in K)\,(c,\, v :=_1 C(c),\, V(c))\,;\\
&(c = n2)_\perp)
\end{aligned}
$$

$n1$ indicates the starting control state, $n2$ indicates the finishing control state, K is a set of possible control states, C a K-indexed family of expressions describing the next control state and V a K-indexed family of expressions specifying the new value of data path v. In a circuit, control states are normally recorded in wires, possibly connected to latches, whereas in a microprocessor this information is typically recorded by a special register known as the program counter.

9.1 Normal form reduction theorems

As in section 7, each high-level programming construct must be handled. As before the finishing control state should be the same as the starting control state for $\texttt{SKIP}$ and there is no change of state of the variables in the program:

$$
\texttt{SKIP} \;\sqsubseteq\; \mathcal{N}(n,\, n,\, \emptyset,\, \emptyset,\, \emptyset)
$$

Typically, the low-level implementation for this would be a single wire labelled by n, taking no time to 'execute'.

For assignment, the control flow is recorded by the C parameter and the new state of the data is recorded by the V parameter. Let $t \geq 1$:

$$
v :=_t e \;\sqsubseteq\; \mathcal{N}(n1,\, n2,\, \{n1\},\, \{n1 \mapsto n2\},\, \{n1 \mapsto e\})
$$

In a clocked hardware implementation, t would indicate the number of clock cycles used to perform the update and the state would be held in a number of latches. A minimum of one clock cycle is required to allow time for the evaluation of the expression e in case this has changed during the previous cycle.

The composition of two circuits already in valid normal form is implemented by taking the union of the K, C and V parameters. The control states in the two circuits must be suitably disjoint. If $K1 \cap K2 = \emptyset$ and $n2 \notin K1$:

$$\mathcal{N}(n1, n, K1, C1, V1) \,; \mathcal{N}(n, n2, K2, C2, V2)$$
$$\sqsubseteq \mathcal{N}(n1, n2, K1 \cup K2, C1 \cup C2, V1 \cup V2)$$

The intermediate control state n is effectively absorbed by the circuit. Typically this would be implemented as an internal wire in the circuit directly connecting the output control state of the first circuit to the input control state of the second one.

A more complete set of theorems for a simple sequential programming language may be found in [27]. A pleasing feature of the approach if mapped to synchronous clocked circuits is that the timing properties of programs are very simple. For example, it is possible to execute all assignment statements in a single clock cycle (if made long enough to accommodate the slowest expression or control state to be evaluated) and for all control constructs to effectively take no clock cycles since control states are determined in parallel to other computation during each clock cycle [44].

Hardware compilation is an active area of research in which we foresee much possible progress, the fusion of mathematical techniques with the practical concerns of engineers, and also an increased potential and desire for hardware/software co-design. The natural parallelism of hardware can be exploited to the full. Many algorithms which would be too slow if implemented using a microprocessor but are too expensive or would be too inflexible if implemented in custom hardware could benefit from such an approach. In addition, the number of levels of abstraction to be handled is reduced, thus helping to decrease the overall possibility for error.

10 Conclusion

This paper has briefly presented a number of approaches investigated by the safemos project to aid the verification of mixed hardware/software systems, especially considering real-time aspects where possible and appropriate. Most of the areas covered are elaborated further in a forthcoming book [7] and other papers referenced here.

A number of formalisms were investigated of the project and not all the parts presented are connected in a satisfactory way. Further work to unify the approaches at different levels of abstraction is certainly required. For example, both HOL and Z were used on the project but the use of

HOL was prevalent because of the desire to mechanise proofs. There is now more support for the mechanisation of proofs in Z [10] so were the project to have started today, more use of Z might have been possible with the same goals of mechanisation in mind.

Whilst the safemos project has now formally finished, further collaborative work on the support of Z using HOL has continued [10] and the related **ProCoS** project continues to aim for the goal of connecting different levels of abstraction in the development process in a mathematical manner. In particular, the approach presented in section 9 looks like a promising one for the development of a provably correct combined hardware/software compiler, thus helping to straddle the development of hardware and software in a unified framework.

Acknowledgements

The research described here was largely supported by the UK Information Engineering Directorate safemos project (IED3/1/1036), which was jointly funded by the UK government DTI (Department of Trade and Industry) and the SERC (Science and Engineering Research Council). We are grateful for the contributions of others involved with the safemos project whose names may be found in [7]. In particular, Mike Gordon (University of Cambridge) and David Shepherd (Inmos) were other key members of the project. Prof. Tony Hoare (Oxford University) and Prof. David May (Inmos) were inspirational in the work undertaken. Finally, John Buckle was an extremely helpful monitoring officer for the project.

Jonathan Bowen is currently funded by the newly formed Engineering and Physical Sciences Research Council (EPSRC) on grant no. GR/J 15186. He Jifeng has been funded by the ESPRIT Basic Research **ProCoS** project and its follow-on (nos. 3104 and 7071).

Bibliography

1. Anceau, F. (1986). *The Architecture of Microprocesors*, Addison-Wesley Publishing Company, Wokingham.

2. Andrews, P.D. *An Introduction to Mathematical Logic and Type Theory: To Truth through Proof*, Computer Science and Applied Mathematics Series, Academic Press.

3. Bowen, J.P. (1990). Formal specification of the ProCoS/safemos instruction set. *Microprocessors and Microsystems, 14*, (10), 631–643.

4. Bowen, J.P. (1992). From programs to object code using logic and logic programming. In R. Giegerich and S.L. Graham, editors, *Code*

Generation – Concepts, Tools, Techniques, Proc. International Workshop on Code Generation, Workshops in Computing, Springer-Verlag, pp.173–192.

5. Bowen, J.P. (1993). Formal methods in safety-critical standards. In *Proc. 1993 Software Engineering Standards Symposium*, IEEE Computer Society Press, pp.168–177.

6. Bowen, J.P. (1993). From programs to object code and back again using logic programming: compilation and decompilation. *Journal of Software Maintenance: Research and Practice*, 5, (4), 205–234.

7. Bowen, J.P. editor (1994). *Towards Verified Systems*, Real-Time Safety Critical Systems Series, Elsevier Science Publishers.

8. Bowen, J.P., et al. (1993). A ProCoS II project description: ESPRIT Basic Research project 7071. *Bulletin of the European Association for Theoretical Computer Science (EATCS)*, *50*, 128–137.

9. Bowen, J.P., Fränzle, M., Olderog, E.-R. and Ravn, A.P. (1993). Developing correct systems. In *Proc. 5th Euromicro Workshop on Real-Time Systems*, IEEE Computer Society Press, pp.176–187.

10. Bowen, J.P. and Gordon, M.J.C. (1994). Z and HOL. In J.P. Bowen and J.A. Hall, editors, *Z User Workshop, Cambridge 1994*, Workshops in Computing, Springer-Verlag, pp.141–167.

11. Bowen, J.P., He Jifeng and Pandya, P.K. (1990). An approach to verifiable compiling specification and prototyping. In P. Deransart and J. Małuszyński, editors, *Programming Language Implementation and Logic Programming*, volume 456 of *Lecture Notes in Computer Science*, Springer-Verlag, pp.45–59.

12. Bowen, J.P. and Stavridou, V. (1993). The industrial take-up of formal methods in safety-critical and other areas: a perspective. In J.C.P. Woodcock and P.G. Larsen, editors, *FME'93: Industrial-Strength Formal Methods*, volume 670 of *Lecture Notes in Computer Science*, Springer-Verlag, pp.183–195.

13. Bowen, J.P. and Stavridou, V. (1993). Safety-critical systems, formal methods and standards. *IEE/BCS Software Engineering Journal*, *8*, (4), 189–209.

14. Brien, S.M. and Nicholls, J.E. (1992). Z base standard. Technical Monograph PRG-107, Oxford University Computing Laboratory, Wolfson Building, Parks Road, Oxford OX1 3QD, UK. Accepted for ISO standardization, ISO/IEC JTC1/SC22.

15. Camilleri, J.A. (1991). Symbolic compilation and execution of programs by proof: A case study in HOL. Technical Report 240, University of Cambridge, Computer Laboratory, UK.

16. Church, A. (1940). A formulation of the simple theory of types. *The Journal of Symbolic Logic*, *5*, 56–68.

17. Good, D.I. and Young, W.D. (1991). Mathematical methods for digital system development. In S. Prehn and W.J. Toetenel, editors, *VDM'91: Formal Software Development Methods, Volume 2*, volume 552 of *Lecture Notes in Computer Science*, Springer-Verlag, pp.406–430.

18. Gordon, M.J.C. (1988). HOL: A proof generating system for Higher-Order Logic. In G. Birtwistle and P.A. Subramanyam, editors, *VLSI Specification, Verification and Synthesis*, Kluwer Academic Publishers, pp.73–128.

19. Gordon, M.J.C. (1991). A formal method for hard real-time programming. In J.M. Morris and R.C. Shaw, editors, *Proc. 4th Refinement Workshop*, Workshops in Computing, Springer-Verlag.

20. Gordon, M.J.C. and Melham, T.F., editors (1993). *Introduction to HOL: A Theorem-proving Environment for Higher-Order Logic*, Cambridge University Press.

21. Gordon, M.J.C., Milner, R. and Wadsworth, C.P. (1979). *Edinburgh LCF: A Mechanised Logic of Computation*, volume 78 of *Lecture Notes in Computer Science*, Springer-Verlag.

22. Hale, R.W.S. (1989). Programming in Temporal Logic. Technical Report 173, University of Cambridge, Computer Laboratory, UK.

23. Hale, R.W.S., Cardell-Oliver, R. and Herbert, J.M.J. (1993). An embedding of Timed Transition Systems in HOL. *Formal Methods in System Design*, *3*, 151–174.

24. Halpern, J., Manna, Z. and Moszkowski, B. (1983). A hardware semantics based on temporal intervals. In *Proc. 10th International Colloquium on Automata, Languages and Programming, Barcelona, Spain*.

25. He Jifeng and Bowen, J.P. (1992). Time interval semantics and implementation of a real-time programming language. In *Proc. 4th Euromicro Workshop on Real-Time Systems*, IEEE Computer Society Press, pp.110–115.

26. He Jifeng, and Bowen, J.P. Specification, verification and prototyping of an optimized compiler. *Formal Aspects of Computing*, to appear.

27. He Jifeng, Page, I. and Bowen, J.P. (1993). Towards a provably correct hardware implementation of Occam. In G.J. Milne and L. Pierre, editors, *Correct Hardware Design and Verification Methods (CHARME'93)*, volume 683 of *Lecture Notes in Computer Science*, Springer-Verlag, pp.214–225.

28. Henzinger, T.A., Manna, Z. and Pnueli, A. (1992). Timed transition systems. In J.W. de Bakker, C. Huizing, W.-P. de Roever, and G. Rozenberg, editors, *Real-Time: Theory in Practice*, volume 600 of *Lecture Notes in Computer Science*, Springer-Verlag.

29. Herbert, J.M.J. (1992). Incremental design and formal verification of microcoded microprocessors. In V. Stavridou, T.F. Melham, and R.T. Boute, editors, *Theorem Provers in Circuit Design*, IFIP Transactions A-10, North-Holland, pp.157–174.

30. Hoare, C.A.R. (1969). An axiomatic basis for computer programming. *Communications of the ACM, 12*, 576–583.

31. Hoare, C.A.R. (1985). *Communicating Sequential Processes*. Prentice Hall International Series in Computer Science.

32. Hoare, C.A.R. (1991). Refinement algebra proves correctness of compiling specifications. In C.C. Morgan and J.C.P. Woodcock, editors, *3rd Refinement Workshop*, Workshops in Computing, Springer-Verlag, pp.33–48.

33. Hoare, C.A.R. and Gordon, M.J.C. editors (1992). *Mechanized Reasoning and Hardware Design*, Prentice Hall International Series in Computer Science.

34. Hoare, C.A.R., He Jifeng, Bowen, J.P. and Pandya, P.K. (1990). An algebraic approach to verifiable compiling specification and prototyping of the ProCoS level 0 programming language. In *ESPRIT '90 Conference Proceedings*, Kluwer Academic Publishers, pp.804–818.

35. Hoare, C.A.R., He Jifeng, and Sampaio, A. (1993). Normal form approach to compiler design. *Acta Informatica, 30*, 701–739.

36. INMOS Limited (1988). *Occam 2 Reference Manual*, Prentice Hall International Series in Computer Science.

37. INMOS Limited (1988). *Transputer Instruction Set: A compiler writer's guide*, Prentice Hall.

38. Krishna Rao, M.R.K., Pandya, P.K. and Shyamasunder, R.K. (1993). Verification tools in the development of provably correct compilers.

In J.C.P. Woodcock and P.G. Larsen, editors, *FME'93: Industrial-Strength Formal Methods*, volume 670 of *Lecture Notes in Computer Science*, Springer-Verlag, pp.442–461.

39. May, D., Barrett, G. and Shepherd, D.E. Designing chips that work. In Hoare and Gordon [33], pp.3–19.

40. Milner, R., Tofte, M. and Harper, R. (1990). *The Definition of Standard ML*, The MIT Press.

41. MoD (1991). The procurement of safety critical software in defence equipment (part 1: Requirements, part 2: Guidance). Interim Defence Standard 00-55, Issue 1, Ministry of Defence, Directorate of Standardization, Kentigern House, 65 Brown Street, Glasgow G2 8EX, UK.

42. Moore, J.S., et al. (1989). Special issue on system verification. *Journal of Automated Reasoning*, *5*, (4), 409–530.

43. Moszkowski, B.C. (1985). A Temporal Logic for multi-level reasoning about hardware. *IEEE Computer*, *18*, (2), pp.10–19.

44. Page, I. and Luk, W. (1991). Compiling Occam into field-programmable gate arrays. In W. Moore and W. Luk, editors, *FPGAs, Oxford Workshop on Field Programmable Logic and Applications*, pages 271–283. Abingdon EE&CS Books, 15 Harcourt Way, Abingdon OX14 1NV, UK.

45. Paulson, L.C. (1987). *Logic and Computation: Interactive Proof with Cambridge LCF*, volume 2 of *Cambridge Tracts in Theoretical Computer Science*, Cambridge University Press.

46. Roscoe, A.W. Occam in the specification and verification of microprocessors. In Hoare and Gordon [33], pp.137–151.

47. Roscoe, A.W. and Hoare, C.A.R. (1988). Laws of Occam programming. *Theoretical Computer Science*, *60*, 177–229.

48. SAFEMOS: Demonstration of the possibility of totally verified systems. Proposal for an IED Research Project (1989). INMOS Ltd, SRI International Cambridge Computer Science Research Center, Oxford University Computing Laboratory and Cambridge University Computer Laboratory.

49. Shepherd, D.E. (1990). Verified microcode design. *Microprocessors and Microsystems*, *14*, (10), 623–630.

50. Spivey, J.M. (1988). *Understanding Z: A Specification Language and its Formal Semantics*, volume 3 of *Cambridge Tracts in Theoretical Computer Science*, Cambridge University Press.

51. Spivey, J.M. (1992). *The Z Notation: A Reference Manual*. Prentice Hall International Series in Computer Science, 2nd edition.

52. SRI International Cambridge Research Center and DSTO Australia. *The HOL System: Description, Tutorial, Libraries, Reference Manual* (1991). Revised version, four volumes.

53. Windley, P.J. (1989). A hierarchical methodology for verifying micro-programmed microprocessors. Technical Report CSE-89-27, University of California, Davis.

Formalising safety in decision support systems

S.K. Das

Artificial Intelligence Group, Department of Computer Science, Queen Mary and Westfield College, University of London

Abstract The safety of a decision support system is a property that ensures that actions recommended by the system will have minimal undesirable consequences. The objective of this paper is to propose a logical formalism for reasoning about safety of a decision support system. A possible world semantics of the proposed logic $\mathcal{L}_{ot}$ is developed and the soundness and completeness result is established. A brief description is given of the implementation of $\mathcal{L}_{ot}$ in Prolog which is motivated by safe task management.

1   Introduction

A *decision support system* [3] is a computerised system which utilizes knowledge about a particular application area to help decision makers by recommending suitable actions. The *safety* of a decision support system [6] is a property that ensures that actions recommended by the system will have minimal undesirable consequences. Undesirable consequences may result from recommendations which have arisen in any of the following situations:

Group I	(a)	Hardware failure. For example, computer hardware, communication hardware, equipment systems hardware, protective system hardware.
	(b)	Human error in the context of operating and maintenance, installation.
Group II	(a)	Incorrect design and specification of the system.
	(b)	Incorrect implementation or one that differs from the actual specification.
Group III	(a)	Inconsistency, redundancy, inaccuracy, and incompleteness of the knowledge base.
	(b)	Incorrect dynamic/static update of the knowledge base.
	(c)	Lack of appropriate *integrity* and *safety constraints* imposed on the system.

Ignoring the hardware failure and human error issues in Group I, for the moment, the so called *logical* or *internal safety* deals with Group II. An important part of internal safety has traditionally been maintained by a rigorous formal approach. Internal safety is always generic in nature and therefore does not need any particular attention for a particular application domain. *Proper* or *external safety* is concerned with Group III. Constraints in this group are collectively called *safety conditions* or *safety knowledge*.

Integrity constraints may be defined as properties which a knowledge base is required to satisfy, for example, a person's age must be less than or equal to 150. Integrity constraints help us to maintain consistent, complete, and accurate knowledge bases. On the other hand, safety constraints are statements involving the notions of obligation and permissibility. For example, a safety related requirement in the context of the action Cisplatin recommendation in the medical domain is "It is obligatory that Cisplatin is stopped if anaphylaxis (severe allergic reaction) occurs". These kinds of safety constraints help us to derive a set of properties and actions which are obligatory at different states of the knowledge base. It is quite important that these obligatory properties and actions be conveyed to the decision makers at appropriate times.

The objective of this paper is to propose a logical formalism for reasoning about the external safety of a decision support system. Our approach is to extend the idea of integrity constraints in database systems [5] to safety knowledge in decision support systems. The proposed logic $\mathcal{L}_{ot}$ takes into account a proper characterization of static and dynamic aspects of the universe of discourse among properties, events, and processes. The language of $\mathcal{L}_{ot}$ enriches the propositional calculus by the introduction of time and the obligation modal operator. Formalisation through modal logic has the major advantage that its possible world semantics reflect the dynamic nature of the world. The generic nature of the logical theory of safety $\mathcal{L}_{ot}$ is intended to ensure its applicability irrespective of any particular application domain.

The rest of the paper is organised as follows. The syntax of the language of $\mathcal{L}_{ot}$ is described in the following section. Section 3 presents axioms and semantics of $\mathcal{L}_{ot}$. The results related to the soundness and completeness of $\mathcal{L}_{ot}$ are stated in this section. The implementation of $\mathcal{L}_{ot}$ is described in section 4. Throughout the rest of the paper, unless otherwise stated, propositional symbols will be denoted by p, q, ..., or by their names such as *Methotrexate*, *IntermittentNausea*, etc.; property symbols by ϕ, ψ, ..., or by their names; action symbols by α, β, ..., or by their names; arithmetic expressions or temporal variables and constants by t, t_1, t_2, ...; formulae by F, G, H, ...; worlds by w, w', ...

2 Syntax

The description of the universe of discourse in $\mathcal{L}_{ot}$ has two aspects: *static* and *dynamic*. The static aspect of the universe of discourse is described by *properties* (e.g., the patient has a cold) and dynamic aspects by *occurrences* (e.g., the patient is given an injection). For a detailed discussion of properties, occurrences, etc. readers are referred to [1]. An occurrence is either an event or a process. Some occurrences involve animate agents (i.e., decision makers) performing *actions*. We shall not consider the cases of occurrences where animate agents do not perform any actions. In light of this discussion, we consider the set of all propositional symbols P is sorted into properties and actions.

We consider the set Z of integers as the constants or time points of the logic. We have the usual arithmetic function symbols. The ordering relation and the equality relation are added to the logic as two special predicate symbols. The domain of propositional formulae is extended to the domain of *formulae* as follows:

- *temporal propositions* $p(t_1, t_2)$ and $p[t_1, t_2]$ are formulae.

- *arithmetic atomic formulae* $t_1 = t_2$ and $t_1 > t_2$ are formulae.

- $[O]F$, $\neg F$, $F \wedge G$ and $\forall t F$ are formulae, where F and G are formulae.

Other logical connectives and the existential quantifier are defined using '$\neg$', '$\wedge$' and '$\forall$' in a usual manner. The inequality $\neq$ and the ordering relation $\leq$ can also be defined using '$=$', '$<$' and logical connectives. A formula is said to be *closed* if it does not have a free variable. In the rest of the paper, we consider only closed formulae.

The temporal proposition $p(t_1, t_2)$ says that if p is a property, then p holds some time during the interval (t_1, t_2) or if p is an action then p is taken some time during (t_1, t_2). In general, we shall say that the proposition p is true in the interval (t_1, t_2). If we know that an action α started at t_1 and finishes at t_2 then $\alpha(t_1, t_2)$ is written as $\alpha[t_1, t_2]$ and $[t_1, t_2]$ will be called an *exact interval*. Similarly, if we know that a property ϕ started holding on or before t_1 and finishes on or after t_2 then $\phi(t_1, t_2)$ is written as $\phi[t_1, t_2]$ and $[t_1, t_2]$ will be called a *compact interval*. The meaning of $[O]F$ is that F is obligatory and somehow to be brought about. We now represent some knowledge base clauses and safety conditions using $\mathcal{L}_{ot}$ syntax.

Following is a safety related requirement in medical protocols for cancer management:

> Following the administration of the drug methotrexate in protocol BO03, it is obligatory that toxicity monitoring is carried out between 24 and 48 hours [8].

$$\forall t_1 \forall t_2 (\text{Methotrexate}[t_1, t_2] \rightarrow [O]\text{ToxicityMonitoring}(t_2 + 24, t_2 + 48))$$

where *Methotrexate* and *ToxicityMonitoring* are events involving actions taken by clinicians and other medical staff. The action *ToxicityMonitoring* is taken some time during $(t_2 + 24, t_2 + 48)$.

Following is an example of a knowledge base clause from the medical domain:

> The intermittent nausea caused by treatment with the drug Cisplatin takes effect within 6 hours and lasts at least 24 hours.

$$\forall t (Cisplatin(t) \rightarrow IntermittentNausea[t + 6, t + 30])$$

where (t) is an abbreviation of (t, t). The duration of the action *Cisplatin* is assumed to be negligible and *IntermittentNausea* is a property of the patient. Since $[t + 6, t + 30]$ is compact, the property *IntermittentNausea* holds in each subinterval of $[t + 6, t + 30]$, for example, $[t + 6, t + 10]$, $[t + 15, t + 16]$, and so on. This means that if the patient suffers from intermittent nausea during $[t + 6, t + 30]$ then the patient also suffers during $[t + 6, t + 10]$, $[t + 15, t + 16]$, and so on.

3   Axioms and semantics

From propositional logic we consider every instance of a propositional tautology as an axiom. These instances of propositional tautologies constructed using temporal propositions and may involve any number of the obligation modal operators, for example, $[O]p(i_1, i_2) \rightarrow [O]p(i_1, i_2)$. We have modus ponens inference rule and substitution rule. We do not require the generalisation rule of inference because the set of formulae considered are closed. We also add the necessary axioms of equality and ordering relations.

If an action is taken in an interval then it is also taken in any interval containing the interval. Similarly, if a property holds in an interval then it also holds in any interval containing the interval. Thus we have the following axiom:

$$\forall t_1 \forall t_2 (p(t_1, t_2) \wedge t_3 \leq t_1 \wedge t_1 \leq t_2 \wedge t_2 \leq t_4 \rightarrow p(t_3, t_4)) \qquad (3.1)$$

If a property holds in a compact interval then it also holds in each of its compact subintervals. In other words, every subinterval of a compact interval is compact and this is axiomatized as follows:

$$\forall t_1 \forall t_2 (\phi[t_1, t_2] \wedge t_1 \leq t_3 \wedge t_3 \leq t_4 \wedge t_4 \leq t_2 \rightarrow \phi[t_3, t_4]) \qquad (3.2)$$

An exact or compact interval is also an interval. This yields the following axiom:

$$\forall t_1 \forall t_2 (p[t_1, t_2] \rightarrow p(t_1, t_2)) \tag{3.3}$$

The system of logic of obligation [2,4] which we have adopted for our purpose has the axiom:

$$\neg [O] \perp \tag{3.4}$$

which represents that nothing impossible is obligatory. Obligations are closed under equivalences as well as distributive over conjunctions. This provides us the following inference rule:

$$\frac{F_1 \wedge ... \wedge F_n \leftrightarrow G}{[O]F_1 \wedge ... \wedge [O]F_n \leftrightarrow [O]G} \tag{3.5}$$

The above axiom will be used in the following context. For our practical purposes we shall be considering knowledge bases consisting of positive Horn clauses of the form $\forall(A_1 \wedge ... \wedge A_n \rightarrow A)$ $(n \geq 0)$ where each A_i and A is a temporal proposition and '$\forall$' is the universal closure. The intended meaning of a knowledge base will be Clarke's completion [5]. A clause is a rule when $n > 0$ and each such rule defines how to execute or bring about an action or a property with the help of other actions or properties. Therefore, the definition of a particular action or a property p will be of the form $\forall(A \leftrightarrow F)$, where A is of the form $p(t_1, t_2)$ or $p[t_1, t_2]$ and F is a disjunction of conjunction of temporal propositions. When we finalise our plan of how to execute or bring about p with the help of a rule, the formula F then becomes the condition of one rule, that is, a conjunction of temporal propositions. Therefore, the formula $\forall(A \leftrightarrow F)$ together with inference rule (3.5) infer what other actions and properties become obligatory when A is obligatory.

A *model* of $\mathcal{L}_{ot}$ is a tuple:

$$< W, V, R >$$

in which W is a set of possible worlds. A world consists of a set of temporally qualified assertions outlining what is known in different periods of time. V is a valuation which associates each pair of world and interval to a subset of the set of propositions. In other words,

$$V : W \times I \rightarrow \Pi(P)$$

where I is the set of all intervals of the form (i_1, i_2) and $[i_1, i_2]$, P is the set of propositions and $\Pi(P)$ is the power set of P. Consider the world w and interval (i_1, i_2) which is a member of I. Then the image of $< w, (i_1, i_2) >$ under the mapping V, written as $V^w(i_1, i_2)$, is the set of all propositions which are true in (i_1, i_2) in the world w. This means that $p(i_1, i_2)$ holds in w for each p in $V^w(i_1, i_2)$. Similarly, we can describe the meaning of $V^w[i_1, i_2]$. R is a serial relation of deontic alternativeness. This relation is relativized to time, that is, $wR_t w'$ means w' is a deontic alternative to w at time t. Note that w' is a deontic alternative to w at time point t only if w and w' are historically identical at t. The presence of axiom $\neg[O]^\perp$ in our system guarantees that R_t is serial.

Given a model $\mathcal{M} = <W, V, R>$, truth values of formulae with respect to pairs $< w, i >$ of worlds and time are determined by the rules given below:

$$\models_{\mathcal{M}}^{<w,i>} p(i_1, i_2) \text{ iff } p \in V^w(i_1, i_2).$$
$$\models_{\mathcal{M}}^{<w,i>} p[i_1, i_2] \text{ iff } p \in V^w[i_1, i_2].$$
$$\models_{\mathcal{M}}^{<w,i>} [O]F \text{ iff for every } w', \text{ if } wR_i w' \text{ then } \models_{\mathcal{M}}^{<w',i>} F.$$

Note that the truth value of a temporal proposition $p[i_1, i_2]$ (or $p(i_1, i_2)$) with respect to a world is same at every time point. But the status of p changes when the current time point lies within the interval $[i_1, i_2]$. A formula F is said to be *true* in model $\mathcal{M}$ if and only if $\models_{\mathcal{M}}^{<w,i>} F$, for every world $< w, i >$ in $W \times I$. A formula F is said to be *valid* if F is true in every model.

Consider the following properties of a valuation V:

$$V^w(i_1, i_2) \subseteq V^w(i_3, i_4), \text{ for every } (i_3, i_4) \text{ containing } (i_1, i_2).$$
$$V^w_\phi[i_1, i_2] \subseteq V^w_\phi[i_3, i_4], \text{ for every } (i_3, i_4) \text{ contained in } (i_1, i_2).$$
$$V^w[i_1, i_2] \subseteq V^w(i_1, i_2), \text{ for every interval } (i_i, i_2).$$

where $V^w_\phi[i_1, i_2]$ is the set of all property symbols of $V^w[i_1, i_2]$. A *standard model* has a valuation structure which satisfies the three properties, known as *supplementation*, *compactness* and *exactness* respectively, for every world. Considering only standard models, we now state the following soundness and completeness results: for every formula $F \in \mathcal{L}_{ot}$, $\models F$ if and only if $\vdash F$.

4   Implementation

In our current Prolog implementation, we have identified three important subsets of $\mathcal{L}_{ot}$ to represent respectively knowledge base, integrity, and safety constraints. A knowledge base clause has the form:

$$\forall(A_1 \wedge \ldots \wedge A_n \rightarrow A) \qquad (n \geq 0)$$

and the general form of an integrity constraint is:

$$\forall(A_1 \wedge \ldots \wedge A_n \rightarrow B_1 \vee \ldots \vee B_m) \qquad (m + n > 0)$$

On the other hand, a safety constraint has the form

$$\forall(A_1 \wedge \ldots \wedge A_n \rightarrow [O]A) \qquad (n \geq 0)$$

In the above three axioms, A is a temporal proposition and each A_i and B_j is either a temporal proposition or an arithmetic atomic formula. The rules and constraints from user inputs are automatically transformed to their equivalent first-order-like internal representation. We have implemented axioms (3.1), (3.2), and (3.3) and inference rule (3.5) as Prolog procedures. Axiom (3.4) is implemented as an integrity constraint.

We have tested the implementation on a number of examples. In the domain of cancer management, we have represented the temporal aspect of a Protocol, TE09, which is concerned with the treatment of testicular cancer [9]. The associated clinical trial comprises activities such as demographic data collection, diagnostic assessment, consideration of eligibility for treatment, randomisation, treatment and follow-up. More formally, we say that trial TE09 is considered completed during an interval say $[t_1, t_2]$ provided the properties demography and eligibility are true at the time point t_1 and that the actions assessment, randomisation, treatment and follow-up are carried out during intervals $[t_1, t_3]$, $[t_3, t_4]$, $[t_4, t_5]$ and $[t_5 + 5, t_2]$ respectively. This corresponds to a rule of the knowledge base. Users input this rule as (<= and & stand for $\rightarrow$ and $\wedge$ respectively):

```
trial_te09[T1,T2] <=
        demography(T1) & eligibility(T1) &
        assessment[T1,T3] & randomisation[T3,T4] &
        treatment[T4,T5] & follow_up[T5+5,T2]
```

Some properties and actions, for example eligibility, in turn are defined by other properties and actions. An example of an integrity constraint is that demography information of a patient must be present in the system at the time of considering the patient's eligibility for the trial. Users input this integrity constraint as (F <= G and G => F are equivalent):

```
eligibility(T) => demography(T)
```

We also have integrity constraints for checking inappropriate values for data entered; for example, obvious restrictions on patients' height and weight (such data is used to calculate an appropriate drug dosage). An example of a safety constraint in TE09 is that it is obligatory that a patient is prehydrated before chemotherapy; otherwise, there is an associated hazard of kidney damage. We represent this as the combination of safety constraint:

```
chemotherapy[T1,T2] => [O]prehydrated(T1)
```

and the rule:

```
prehydration[T1,T2] => prehydrated(T2)
```

where `chemotherapy` and `prehydration` are actions and `prehydrated` is a property. Integrity and safety constraints are verified upon each update to the knowledge base and at every successive time increment. Recall that in our formalisation, the set of propositional symbols has been divided into actions and properties. As time changes, properties change their state from true to false and vice versa. Similarly, actions change their state from active to complete, active to suspended, pending to complete, and so on. We have therefore devised two sets of states of actions and properties during the course of their execution.

The possible states for an action are active, pending, complete, remind, suspended and aborted. An action is active if it is being executed, that is, the current time point lies after the starting point of execution of the action and the action is still being executed. If an action is defined by other actions then as long as one of its child actions is active, the action itself is active. An action is complete if each of its child actions is complete. In the case of a complete action, the current time point is beyond the finishing point of the action. An action is pending if the current time point lies beyond the time point where it is supposed to start execution in the plan. An action is in reminding status if the current time point lies before the time point where it is supposed to start execution in the plan. A suspended action has been started but is no longer active in the plan. The time taken by an aborted action is less than the minimum time it should take under any circumstances. This minimum time for an action is supplied in the knowledge base. It can be obligatory to perform an action within a certain time interval. The state of an obligatory action is one of active, pending, suspended, or aborted.

The status of a property is one of true, false, or pending with respect to the current time point. A property is true if the current time point lies within an interval where the property is true. Similarly, a property is false if the current time point lies outside each interval where the property is true. A property may be defined by other properties and occurrences. In such cases, a property is true if each of its child properties or occurrences is true and false if one of its child properties or occurrences is false. Pending properties are defined in a similar manner. It can be obligatory to bring about a property within a certain time interval. The state of an obligatory property is one of true, false, or pending.

To describe in detail how the task management is carried out safely by the implementation, we consider a small, artificial, but complete knowledge base consisting of the following rules:

```
task_01[T1,T2] <= prop(T1) & task_11[T1,T3] &
                  task_12[T1,T3] & task_13[T3,T2]
task_11[T1,T2] <= task_21[T1,T2]
task_11[T1,T2] <= task_22[T1,T2]
```

where **prop** is a property and **task_01**, **task_11**, **task_12**, **task_13**,
task_21, and **task_22** are actions with minimum durations 12, 3, 3, 9, 3,
and 3 respectively. The first rule says that **task_01** is said to be completed
within an interval [T1,T2] if the property **prop** is true at T1 and the tasks
task_11, **task_12**, and **task_13** are completed within [T1,T3], [T1,T3],
and [T3,T2]. The other two rules are interpreted in a similar fashion.
Obviously, **task_11** can be completed in two different manners using the
second or third rules. As part of safety knowledge, we have an integrity
constraint:

```
task_21[T1,T2] => prop[T1,T2]
```

which says that **prop** must be true throughout the execution of **task_21**;
otherwise, it does not make sense of **task_21**. We also have a safety constraint:

```
task_12[T1,T2] => [O]task_13(T2,T2+10)
```

whose interpretation is that it is obligatory to complete **task_13** within
10 units of time after **task_12** is completed. We have an interface to
manipulate the knowledge base and another interface to display the status
of properties and actions in terms of their state of completion, etc. Initially,
the total plan is displayed by the latter interface as follows:

```
          <-*..1..2..3..4..5..6..7..8..9..0..1..2..3..4..->
task_01       [...................................]PENDING
 prop         []PENDING
 task_11      [........]PENDING
  task_21     [........]PENDING

  task_22     [........]PENDING
 task_12      [........]PENDING
 task_13               [........................]REMIND


              ----------------
              automatic    (a)
              branching    (b)
              node input   (n)
              query        (q)
              redraw plan  (r)
              time update  (t) -> currently(0)
              adjust view  (v) -> currently(0,0)
              exit         (x)
              ----------------
              option?
```

The current time is denoted by the asterisk mark on the time line (top
row) and this can be incremented by selecting the option 'time update'.
The second occurrences of 0, 1, ..., 9 in the time line are actually 10, 11, ...,
19 respectively, and so on. If the displaying screen cannot accommodate
the total plan of a particular application then any part of the plan can be
displayed by selecting the option 'adjust view'. The two different ways to
complete task_11 are separated vertically by a space. Suppose we choose
to complete task_11 using task_21. We confirm this to the system by
selecting the option 'branching' and the final plan is displayed by the system
as follows:

```
          <-*..1..2..3..4..5..6..7..8..9..0..1..2..3..4..->
task_01       [...................................]PENDING
 prop         []PENDING
 task_11      [........]PENDING
  task_21     [.......]PENDING
 task_12      [........]PENDING
 task_13               [........................]REMIND
```

It is not necessary to finalise the plan beforehand. Alternative options
can be selected to complete a particular task as they are required. The
tasks which we are supposed to start at this time point are pending and

the tasks which we should start in the future are at reminding status. The current time is set at zero. If we try to input `task_21[0,3]` using the option 'node input', the integrity constraint violation message will be displayed as `prop` is not true yet, contravening the integrity constraint. We input first the fact `prop[0,3]` and then `task_21[0,3]`. The former states that `prop` is true at the current time and is expected to be true during the execution of `task_21`. The fact `task_21[0,3]` states that we hoping to complete `task_21` and therefore `task_11` in `[0,3]`. The integrity constraint is now satisfied. We also input `task_12[0,3]` at this time point. The status of the actions and properties look as follows at time 0:

```
              <-*..1..2..3..4..5..6..7..8..9..0..1..2..3..4..->
task_01       [.......................................]ACTIVE
 prop         [........]TRUE
 task_11      [........]ACTIVE
   task_21    [........]ACTIVE
 task_12      [........]ACTIVE
 task_13            [...............................]REMIND[O]
```

Due to the presence of `task_12[0,3]`, the safety constraint will derive `[O]task_13[3,13]` which means `task_13` is now obligatory in `[3,13]`. This has been highlighted by an extra '[O]' beside the reminding status of `task_13`. After the current time point is advanced by four units, the status of the tasks now takes the following form:

```
              <-0..1..2..3..*..5..6..7..8..9..0..1..2..3..4..->
task_01       [.......................................]PENDING
 prop         [........]FALSE
 task_11      [........]COMPLETE
   task_21    [........]COMPLETE
 task_12      [........]COMPLETE
 task_13            [...............................]PENDING[O]
```

At this time point, `prop` is false, `task_21` and hence `task_11` is complete as well as `task_12`; `task_13` is now pending because it has not been started yet. In addition `task_13` is still tagged with '[O]' because it is still obligatory. We input `task_13[3,13]` to start `task_13` at time point 3 and complete at time point 13. The status of actions and properties now look like the following:

```
              <-0..1..2..3..*..5..6..7..8..9..0..1..2..3..4..->
task_01       [.................................................]ACTIVE
 prop         [........]FALSE
  task_11     [........]COMPLETE
   task_21    [........]COMPLETE
  task_12     [........]COMPLETE
  task_13              [...........................]ACTIVE
```

The status of `task_13` is only active and does not contain '[O]' since according to the input provided to the system we can complete `task_13` in [3, 13] (that is, also in (3, 13), by axiom (3.3)) to satisfy the safety condition. At time point 12, we update `task_13[3,13]` by `task_13[3,12]` to inform the system that we have completed `task_13` in exactly nine units of time which is one unit before we planned. Note that completing `task_13` in the interval [3,12] also implies completing the same in the interval (3,13) (axiom (3.1)). Finally, at time point 13, the total history for the completion of `task_01` looks like the following:

```
              <-0..1..2..3..4..5..6..7..8..9..0..1..2..*..4..->
task_01       [.................................................]COMPLETE
 prop         [........]FALSE
  task_11     [........]COMPLETE
   task_21    [........]COMPLETE
  task_12     [........]COMPLETE
  task_13              [...........................]COMPLETE
```

At this state of the knowledge base a query of the form `task_01[T1,T2]` will return instantiations of `[T1,T2]` to `[0,12]`. But a query of the form `task_01(T1,T2)` will return instantiations of `(T1,T2)` to `(0,12)` and each interval containing `(0,12)` due to axiom (3.1). On the other hand, a query of the form `prop[T1,T2]` will return instantiations of `[T1,T2]` to `[0,3]` and each of its subintervals.

5   Conclusion

We have developed a logic $\mathcal{L}_{ot}$ for reasoning about external safety in decision support systems. The quantifiers, arithmetic functions, and binary predicate symbols have been introduced into the logic $\mathcal{L}_{ot}$ just to manipulate the intervals. But the application specific symbols such as *Cisplatin*, *OpenTap* have been considered as propositional. The logic can be extended to full first-order. For that we need to consider the underlying set of constants as sorted. This is because the temporal constants are of type integer. The main intended application of $\mathcal{L}_{ot}$ is to reason about safety of a medical decision support system [7] although the logic is generic enough to be used in other areas of artificial intelligence wherever the concepts of action, obligation, and time occur. Our implementation is essentially a prototype and we have provided a scheme for deontic and temporal reasoning rather than a robust implementation.

Acknowledgements

The author would like to thank his colleagues in the RED project, especially Peter Hammond and John Fox of the Imperial Cancer Research Fund for many helpful discussions on this work. The author is supported under the DTI/SERC project ITD 4/1/9053: Safety-Critical Systems Initiative.

Bibliography

1. Allen, J.F. (1984). Towards a general theory of action and time. *Artificial Intelligence*, *23*, 123–154, North-Holland.

2. Aqvist, L. (1984). Deontic Logic. *Handbook of Philosophical Logic*, Editor: D. Reidel, *2*, 605–714.

3. Bonczek, R.H., Holsapple, C.W. and Whinston, A.B. (1984). Development in decision support systems. *Advances in Computers, 3*, Academic Press, 141-175.

4. Chellas, B. (1980). *Modal Logic*. Cambridge University Press.

5. Das, S.K. (1992). *Deductive Databases and Logic Programming*, Addison-Wesley.

6. Das, S.K. and Fox, J. (1993). A logic for reasoning about safety in decision support systems. *Proceeding of the second European Conference on Symbolic and Quantitative Approaches to Reasoning and Uncertainty*, Springer-Verlag.

7. Hammond, P., Harris, A.L., Das, S.K. and Wyatt, J.C. (1994). Safety and decision support in oncology. *Methods of Information in Medicine*, in press.

8. MRC. (1986). A randomised trial of two chemotherapy regimens in the treatment of operable osteosarcoma. *MRC Protocol BO03/EORTC 80861.*

9. MRC. (1990). Carboplatin based combination chemotherapy for good prognosis metastatic malignant teratoma. MRC Cancer Trial Office, Cambridge.

Formal techniques for requirements analysis for safety-critical systems

R. de Lemos[1&2]**, A. Saeed**[3] **and T. Anderson**[1&3]

[1]*Department of Computing Science,* [2]*Department of Chemical and Process Engineering and* [3]*BAe Dependable Computing Systems Centre, University of Newcastle upon Tyne*

Abstract Formal support for the different activities performed during requirements analysis demands the utilisation of a set of formal notations and techniques whose features and expressive power match the characteristics of the activities. Selecting an appropriate formal technique for an activity allows emphasis to be placed on pertinent characteristics of the system, enabling the technique to work to its own strengths. In order to facilitate the utilization of different formal techniques, in this paper, we introduce an event/action model (E/A model) as a common foundation for models of system behaviour. To show the flexibility of the E/A model, we incorporate its concepts into two different classes of formalisms.

Keywords: safety-critical systems, requirements analysis, timeliness requirements, formal techniques, property-oriented formalisms, operational formalisms.

1 Introduction

For formal support to be effective in system development it is essential to examine the demands imposed by the context in which application is envisaged. If formal support is considered in isolation, more emphasis may be placed on the mathematical properties of a formal notation than a method to guide its application, rendering the support impractical. In this paper, we focus on the provision of formal support for the requirements stage of system development, for the class of process control systems.

In addition to the usual features that should be provided by formal techniques with respect to their application in requirements analysis, such as an unambiguous notation, checks for consistency and completeness, a number of specific features arise from the class of systems under consideration.

1. Compatibility with the underlying models of control theory, such as differential equations and variables that are functions of time.

2. Support traceability (by formal refinement and verification) between the results of system safety techniques, such as hazard identification and software specifications.

3. Promote the analysis of distinct properties of system behaviour from different perspectives (e.g. timing and reliability).

During requirements analysis, the broad range of information that must be encoded and analysed suggests two alternatives for the basis of formal support: the employment of a single *wide spectrum notation*, such as Duration Calculus [1], [2] or extended CCS [3]; or a number of *specialized notations*, such as RTL [4] and Statecharts [5]. In the case of wide spectrum notations features 1 and 2 are supported, in the sense that it is possible with appropriate extensions to relate to the models of control theory and relate the results of system safety analysis to software specifications. However, because of the many interacting features such notations tend to be complex making it difficult to extract a suitable subset of the notation for specific analysis, working against feature 3. On the other hand, a suitable set of specialized notations will support features 1 and 3, in the sense that an appropriate formalism can be selected for a related class of properties, and the different notations permit a selective approach to the analysis of the requirements. However, an inconvenience of this approach is that difficulties arise when attempting to link specifications expressed in different formalisms which works against feature 2.

In this paper, we follow an approach for formal support based on specialized notations which enables the analyst to select the appropriate technique in accordance with the properties of the system to be analyzed. To support this approach we have proposed a framework which provides the rationale for the utilization of different techniques [6], and an event/action model which provides the common foundation for the models of the system behaviour.

The framework, which has the purpose to facilitate the systematic analysis of the requirements, has a structure which follows from the analysis of the system that identifies the key components (recursively, a component can be considered to be another system) and their interactions (definition of the interface between the components and the behaviour observed at that interface), establishing the domains of analysis and their inter-relationships, respectively. This process is conducted recursively, each decomposition leading to a lower level of abstraction. The framework is defined by associating its phases with the domains of analysis, and the ordering of the phases with the identified inter-relationships. At each phase of the framework formal notations and techniques are used to represent and reason about the behaviour and properties of a system. Although formal techniques have been classified in a number of different ways, in the context of our proposed approach two classes of formalisms are identified: *property-oriented* and *operational*. The degree of application of one class of formalism versus the other is related to the level of abstraction being considered: at higher levels of abstraction there is a natural tendency to use property-oriented formalisms, whereas at lower levels operational formalisms dominate.

The event/action model (E/A model) describes the behaviour of real-time safety-critical systems, which exhibit continuous and discrete behaviours, in both value and time domain. The E/A model provides a set of primitive concepts which enable the modelling of the system behaviour, namely, events, actions and states, and the concept of a time structure (or timeline). In the approach taken, those variables which are continuous have their behaviour discretised according to imposed thresholds and discontinuities that they are subjected to.

This paper supplements the earlier work on the E/A model [7], [8] by providing a formal interpretation for the E/A model concepts and a more precise treatment of the PEA notation. In [7] the E/A model was initially introduced and its concepts defined only from the time domain perspective; one of the aims of the E/A model was to provide a formal interpretation of timing diagrams. In [8] the PEA notation was introduced as a compact notation to facilitate the formal analysis of system behaviour in terms of the E/A model concepts, in both value and time domain; through two case studies we emphasized the usefulness in applying the E/A model concepts at the requirements stage.

The rest of the paper is presented as follows. In section 2, we define the syntax and semantics of the E/A model, and its corresponding PEA notation, to be employed in the description of system behaviour. Section 3 presents an extract of a case study and its analysis from the perspective of the PEA notation. Section 4 describes how the E/A model is represented in property-oriented and operational formalisms, and illustrates their application on the case study. Finally, section 5 contributes with some concluding remarks.

2 Behaviour description

2.1 Event/Action model (E/A model)

The E/A model is based on the following primitive concepts. A *state* of a system is the information that, together with the system input, determines the behaviour of the system. A *transition* represents a transformation in the system state. The system state is modified by the occurrence of events and the execution of actions. An *event* is a temporal marker of no duration which causes or marks a transition. An *action* is the basic unit of activity which implies duration. The *duration* of an interval is the time distance between the two events that define the interval. Apart from events, actions and states which describe the behaviour of process control systems, the E/A model also takes into account the timing uncertainties associated with them.

The motivation for selecting these primitive concepts is twofold: they have been used as primitives in several real-time specification languages [9], [4], and they have meaningful interpretations at different levels of abstraction. These concepts provide flexibility, enabling descriptions to be given of system behaviour ranging from the activities of the physical entities of the plant to the temporal ordering of the computational tasks of the control system. The main features of the E/A model are: the primitive concepts can be expressed in different classes of formalisms, both discrete and dense time structures are supported, and timing constraints can be depicted graphically.

2.1.1   Definition of the E/A model

The E/A model is defined in terms of an extended first order logic that includes variables and predicates that are interpreted as functions from a time domain. The predicates of the extended first order logic are also referred to as state predicates and those predicates that are used to specify a transition (i.e. mark a time point) referred to as transition predicates. In the following we present the syntax and semantics of the E/A model.

Syntax

The underlying model consists of a time independent and a time dependent part. The time independent part consists of a set of variable names, a set of predicate names and a set of function names. Each variable name denotes a value from an associated type, each predicate name a predicate with a fixed arity n and associated domain, and each function name a function with a fixed arity n and the associated domain and range. The time dependent part consists of a set of time dependent variables and a set of time dependent predicates. Each variable name denotes a function from the time domain to a type, each predicate name denotes a function from the time domain to a predicate. The passage of time (time domain) is represented by a time structure $\mathbf{T}$ which is isomorphic to the non-negative reals or a subset of the them (e.g. natural numbers).

For the above model, a term is either a variable or a n-ary function evaluated over n terms, and a predicate is defined inductively as:

1. A well-defined expression over n terms using the relational operators (e.g. $<$ and $\leq$), a time independent predicate or a time dependent predicate is an atomic predicate.

2. If p and q are predicates, so are $\neg p$ and $p \wedge q$.

3. If p is a predicate and x is a time independent variable of type V_x, $\forall x \in V_x : p$ is a predicate.

All time dependent variables and predicates are piecewise continuous (i.e. there are a finite number of discontinuities and limits exist from both the left and right at each discontinuity). Predicates satisfy the *finite variability* property, since a predicate can only change at a discontinuity.

We associate *instance number* to a predicate to specify the number of times for which the predicate has been true. Instance numbers are a convenient way to describe the behaviour of discrete variables, or continuous variables that are discretised, because references to the past and future behaviours of predicates are allowed.

In order to express the instance number of a state predicate we introduce the following notation $sp(t)^i$, where $sp(t)$ is a state predicate.

The first instance of a state predicate is defined as follows:

$$\forall t \in \mathbf{T} : [sp(t)^1 \iff sp(t) \wedge (\exists t_1 \in \mathbf{T} : t_1 \leq t \wedge (\forall t_2 \in \mathbf{T} : t_2 < t_1 \Rightarrow$$

$$\neg sp(t_2)) \wedge (\forall t_3 \in \mathbf{T} : t_1 \leq t_3 \leq t \Rightarrow sp(t_3)))].$$

The ith instance of a state predicate, represented by the index $i(i > 1)$, is defined as follows:

$$\forall i \in \mathbf{I}^+ : i > 1 \Rightarrow$$

$$\forall t \in \mathbf{T} : [sp(t)^i \iff sp(t) \wedge \exists t_1, t_2, t_3 \in \mathbf{T} : t_1 < t_2 < t_3 < t \wedge$$

$$\forall t_4 \in \mathbf{T} : t_1 \leq t_4 < t_2 \Rightarrow sp(t_4)^{i-1} \wedge \forall t_5 \in \mathbf{T} : t_2 \leq t_5 < t_3 \Rightarrow \neg sp(t_5) \wedge$$

$$\forall t_6 \in \mathbf{T} : t_3 \leq t_6 \leq t \Rightarrow sp(t_6)].$$

In order to capture a *transition predicate* we introduce the bar operator "|". Figure 1 depicts the possible transition predicates that can be obtained from a state predicate. By applying the operator "$|^C$" to a state predicate we capture the first point in time at which a state predicate becomes **true** or **false**. This transition, known as a *closed transition*, is defined as follows:

$$\forall t \in T. \, [\rvert^{o}(sp(t)) \Leftrightarrow \neg sp(t) \wedge$$

$$(\exists \delta \in T. \, \forall t_1 \in T. \, t - \delta < t_1 < t \Rightarrow \neg sp(t_1) \wedge \forall t_2 \in T. \, t < t_2 < t + \delta \Rightarrow sp(t_2)$$

$\rvert^{C}(sp(t))$ T   F   t $\rvert^{o}(sp(t))$ T   F   t

$\rvert^{C}(\neg sp(t))$ T   F   t $\rvert^{o}(\neg sp(t))$ T   F   t

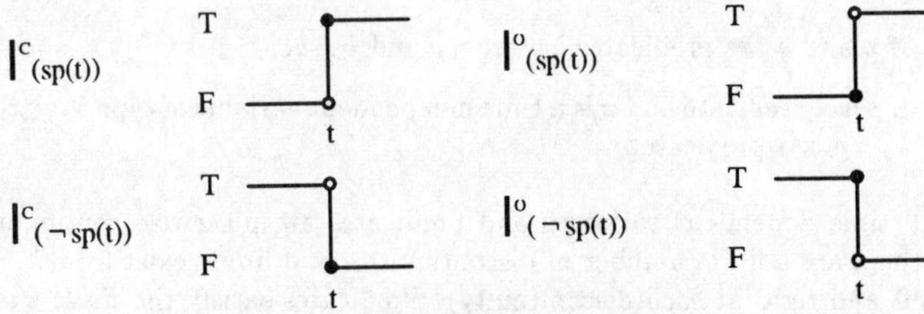

Figure 1. Transitions over state predicates

$$\forall t \in \mathbf{T} : [\rvert^{C}(sp(t)) \Longleftrightarrow sp(t) \wedge$$

$$(\exists \delta \in \mathbf{T} : \forall t_1 \in \mathbf{T} : t - \delta < t_1 < t \Rightarrow \neg sp(t_1) \wedge$$

$$\forall t_2 \in \mathbf{T} : t < t_2 < t + \delta \Rightarrow sp(t_2))].$$

The value of δ is defined between the time separation (distance) between two consecutive observations (or samplings) of the value of the variables, used to construct the predicate. For continuous variables, δ has to be defined as a ratio of the highest frequency of the variable (or the Nyquist frequency), and for discrete variables, δ has to be defined in terms of the smallest time period for which the variable remains stable [10].

By applying the operator "$\rvert^{O}$" to a state predicate we capture the last point in time just before a predicate becomes true or false. This transition, known as *open transition*, is defined as follows:

$$\forall t \in \mathbf{T} : [\rvert^{O}(sp(t)) \Longleftrightarrow \neg sp(t) \wedge$$

$$(\exists \delta \in \mathbf{T} : \forall t_1 \in \mathbf{T} : t - \delta < t_1 < t \Rightarrow \neg sp(t_1) \wedge$$

$$\forall t_2 \in \mathbf{T} : t < t_2 < t + \delta \Rightarrow sp(t_2))].$$

In order to capture the instance number of a transition, we introduce the bar operator with an index i added to it "$\rvert_i$". The first instance of a closed transition is defined as follows:

$$\forall t \in \mathbf{T} : [\rvert^{C}_{1}(sp(t)) \Longleftrightarrow \rvert^{C}(sp(t)) \wedge (\forall t_1 \in \mathbf{T} : t_1 < t \Rightarrow \neg \rvert^{C}(sp(t_1))].$$

The ith instance of a closed transition, represented by the index $i (i > 1)$, is defined as follows:

$$\forall t \in \mathbf{T} : \forall i \in \mathbf{I}^+ : [i > 1 \Rightarrow |_i^C(sp(t)) \Longleftrightarrow |^C(sp(t)) \wedge$$

$$(\exists t_1 \in \mathbf{T} : t_1 < t \wedge |_{i-1}^C(sp(t_1)) \wedge$$

$$(\forall t_2 \in \mathbf{T} : t_1 < t_2 < t \Rightarrow \neg|^C(sp(t_2))))].$$

The definition of the single bar operator for the open transition is similar to the one for the closed transition. In the sequel, unless otherwise mentioned, a bar operator without a superscript will refer to a closed transition.

Semantics

For the time independent part, the valuation functions $\mathcal{V}$, $\mathcal{P}$ and $\mathcal{F}$ are respectively defined for the variables, predicates and functions in the traditional way as for a first-order logic. For the time dependent part the valuation functions $\mathcal{V}_T$ and $\mathcal{P}_T$ are defined for the variables and predicates as follows:

for a variable v_i with the range $V_{v_i} - \mathcal{V}_T(v_i) : \mathbf{T} \rightarrow V_{v_i}$

for a predicate p_i with range V_j for term $tr_j - \mathcal{P}_T(p_i(tr_1, \cdots, tr_n)) :$
$\mathbf{T} \rightarrow V_1 \times \cdots \times V_n \rightarrow \mathbf{B}$.

The interpretation $\mathcal{J}$ for the underlying model is defined by the structure $(\mathcal{V}, \mathcal{P}, \mathcal{F}, \mathcal{V}_T, \mathcal{P}_T)$, and is defined pointwise.

The interpretation of a term tr is defined as follows:

$\mathcal{J}(tr)(t) = \mathcal{V}(tr)$, if tr is a time independent variable

$\mathcal{J}(tr)(t) = \mathcal{V}_T(tr)(t)$, if tr is a time independent variable

$\mathcal{J}(tr)(t) = \mathcal{F}(f_i)(\mathcal{J}(tr_1)(t), \cdots, \mathcal{J}(tr_n)(t))$, if tr is a function $f_i(tr_1, \cdots, tr_n)$.

The interpretation of the basic predicates is defined as follows:

$\mathcal{J}(p(tr_1, \cdots, tr_n))(t) = \mathcal{P}(p)(\mathcal{J}(tr_1)(t), \cdots, \mathcal{J}(tr_n)(t))$,
if p is a time independent predicate

$\mathcal{J}(p(tr_1, \cdots, tr_n))(t) = \mathcal{P}_T(p)(t)(\mathcal{J}(tr_1)(t), \cdots, \mathcal{J}(tr_n)(t))$,
if p is a time dependent predicate.

70   *R. de Lemos et al.*

For a given interpretation $\mathcal{J}$, each predicate p is evaluated by the function $\mathcal{J}(p) : \mathbf{T} \to \mathbf{B}$.

$$\mathcal{J}(\neg p)(t) = \text{true iff } \mathcal{J}(p)(t) = \text{false}$$

$$\mathcal{J}(p \wedge q)(t) = \text{true iff } \mathcal{J}(p)(t) = \text{true and } \mathcal{J}(q)(t) = \text{true}$$

$$\mathcal{J}(\forall x \in V_x : p)(t) = \text{true iff } \mathcal{J}'(p)(t) = \text{true}$$

for every $\mathcal{J}'$ that has the same valuation functions as $\mathcal{J}$ except for $\mathcal{V}'(y)$ which is the same as $\mathcal{V}(y)$ for every time independent variable not equal to x.

2.1.2   Primitive functions of the E/A model

The primitive concepts of the E/A model are related to the timeline by two types of primitive functions: point and interval. A *point function* is a temporal marker of no duration, represented as a cut in the timeline, which models events. An *interval function* denotes a duration, represented as a contiguous section of the timeline, which models states and actions. The timing uncertainties associated with point and interval functions can be represented in terms of a *utility function*. The primitive functions have time and instance number as parameters.

In the definition of the primitive functions, conditions (state and transition predicates) capture the value domain properties of the functions, and time points capture the time domain properties of the functions. In order to distinguish the value domain from the time domain definitions, the subscript "V" or "T", respectively, is added to the name of the function. The definitions of the primitive functions, in the value domain, will be made only in terms of the closed transition. As a consequence, the upper and lower boundaries of a time interval will be respectively closed ("[") and open (")"), however other combinations could be employed depending on the type of transitions being considered.

Point function

A point function $E(t, i)$ is a function which maps the timeline ($\mathbf{T}$ the set of all of its points) and the number of instances of an event ($\mathbf{I}^+$ the set of the nonnegative integer numbers) into a Boolean.

In the value domain the definition of the point function is in terms of the ith instance of a condition which is specified by a transition predicate:

$$E_V(t, i) : \mathbf{T} \times \mathbf{I}^+ \to \mathbf{B}$$

$$\forall t \in \mathbf{T} : \forall i \in \mathbf{I}^+ : [E_V(t, i) \Longleftrightarrow |_i(sp_E(t))].$$

In the time domain the definition of the point function is in terms of the time point constants t_E^i which mark the ith instance of the event:

$$E_T(t, i) : \mathbf{T} \times \mathbf{I}^+ \to \mathbf{B}$$

$$\forall t \in \mathbf{T} : \forall i \in \mathbf{I}^+ : [E_T(t, i) \Longleftrightarrow t = t_E^i] \quad t_E^i \in \mathbf{T}_E = \{t_E^1, t_E^2, \cdots\}.$$

Instead of the notation (t_E^i), the occurrence function $(@(E, i))$ from RTL could have been employed [4].

The value and time domain definitions of a point function are alternative forms for describing the occurrence of the same event:

$$\forall t \in \mathbf{T} : \forall i \in \mathbf{I}^+ : [E_V(t, i) \Longleftrightarrow E_T(t, i)].$$

A relative time representation of the timeliness requirements of a sequence of events can impose one of three basic types of timing constraints [11]:

minimum   -   no less than t time units must elapse between the occurrence of two events;

maximum   -   no more than t time units must elapse between the occurrence of two events;

durational   -   exactly t time units must elapse between the occurrence of two events.

In a similar way as presented for RTL [4], in the following we state two *monotonicity* properties which are associated with the point function:

uniqueness property   - at most one time point can be associated with each occurrence of an event, i.e. the same instance number of an event cannot happen at two distinct time points:

$$\forall t, t' \in \mathbf{T} : \forall i \in \mathbf{I}^+ : [E(t, i) \wedge E(t', i) \to t = t'].$$

ordering property   - if the ith occurrence of an event happens, then the previous occurrences of the same event must have happened earlier, hence two distinct occurrences of the same event must happen at different time points:

$$\forall t, t' \in \mathbf{T} : \forall i, j \in \mathbf{I}^+ : [E(t, i) \wedge E(t', j) \wedge i < j \to t < t'].$$

For discrete timelines, in order to observe the occurrence of all instances of the same event, the granularity of the timeline should be smaller than the smallest time interval between the occurrence of any two events. If the same event occurs more than once between two ticks of the timeline, only one of the occurrences will be observed.

Interval function

An *interval function* $A(t, i)$ is a function which maps the timeline and the number of instances of an action (or a state) into a Boolean. (In the following, the definition of the interval function will be restricted to the execution of an action, however it could be extended to the holding of states.) An action is manifested in the system by its associated events: the *start event* that marks the initiation of an action, and the *finish event* that marks the completion of an action.

In the value domain, the interval function is defined in terms of the transition predicates corresponding to the start and finish conditions, and the state predicate corresponding to invariant condition. This is represented as follows:

$$A_V(t) : \mathbf{T} \to \mathbf{B}$$

$$\forall t \in \mathbf{T} : [A_V(t) \iff \exists t_1, t_2 \in \mathbf{T} : t_1 \leq t_2 \wedge (\mathsf{I}(sp_{\uparrow A}(t_1)) \wedge \mathsf{I}(sp_{\downarrow A}(t_2)) \wedge$$

$$\forall t_3 \in \mathbf{T} : t_1 \leq t_3 < t_2 \Rightarrow sp_{INV}(t_3) \wedge (\neg sp_{\downarrow A}(t_3))].$$

The first instance of an interval function is defined as follows:

$$A_V(t, 1) : \mathbf{T} \times \mathbf{I}^+ \to \mathbf{B}$$

$$\forall t \in \mathbf{T} : [A_V(t, 1) \iff A_V(t) \wedge (\exists t_1, \in \mathbf{T} : t_1 \leq t) \wedge$$

$$(\forall t_2 \in \mathbf{T} : t_2 \leq t_1 \Rightarrow \neg A_V(t_2)) \wedge (\forall t_3 \in \mathbf{T} : t_1 \leq t_3 \leq t \Rightarrow A_V(t_3))].$$

The ith instance of an interval function is defined as follows:

$$A_V(t,i) : \mathbf{T} \times \mathbf{I}^+ \to \mathbf{B}$$

$$\forall i \in \mathbf{I}^+ : i > 1 \Rightarrow$$

$$\forall t \in \mathbf{T} : [A_V(t,i) \iff A_V(t) \wedge \exists t_1, t_2, t_3 \in \mathbf{T} : t_1 < t_2 < t_3 < t \wedge$$

$$\forall t_4 \in \mathbf{T} : t_1 \leq t_4 < t_2 \Rightarrow A_V(t_4, i-1) \wedge \forall t_5 \in \mathbf{T} : t_2 \leq t_5 < t_3 \Rightarrow$$

$$\neg A_V(t_5) \wedge \forall t_6 \in \mathbf{T} : t_3 \leq t_6 \leq t \Rightarrow A_V(t_6)].$$

In the value domain, the start and finish events are defined, respectively, as the *start condition* and the *finish condition*. The start condition defines the condition that triggers the execution of an action, it is defined as follows:

$$\uparrow A_V(t,i) : \mathbf{T} \times \mathbf{I}^+ \to \mathbf{B}$$

$$\forall t \in \mathbf{T} : [\uparrow A_V(t,i) \iff |(A_V(t,i))].$$

The finish condition defines the condition in which an action terminates its execution, it is defined as follows:

$$\downarrow A_V(t,i) : \mathbf{T} \times \mathbf{I}^+ \to \mathbf{B}$$

$$\forall t \in \mathbf{T} : [\downarrow A_V(t,i) \iff |(\neg A_V(t,i))].$$

In the time domain, the start and finish events are defined as temporal markers, respectively, the *start time* and the *finish time*. The definition of the start and finish times follows directly from the time domain definition of the point function. They are defined in terms of time point constants of the form $t^i_{\uparrow A}$ and $t^i_{\downarrow A}$ at which the respective events $\uparrow A_T(t,i)$ and $\downarrow A_T(t,i)$ have occurred for the ith time.

In the time domain, the interval function $A_T(t,i)$ is defined, as follows, in terms of the start and finish times of an action:

$$A_T(t,i) : \mathbf{T} \times \mathbf{I}^+ \to \mathbf{B}$$

$$\forall t \in \mathbf{T} : \forall i \in \mathbf{I}^+ : [A_T(t,i) \iff t^i_{\uparrow A} \leq t < t^i_{\downarrow A}].$$

The *start time* refers to the time point at which the start event of an action occurs:

$$\uparrow A_T(t,i) : \mathbf{T} \times \mathbf{I}^+ \to \mathbf{B}$$

$$\forall t \in \mathbf{T} : \forall i \in \mathbf{I}^+ : [\uparrow A_T(t,i) \iff t = t^i_{\uparrow A}]$$

$$t^i_{\uparrow A} \in \mathbf{T}_{\uparrow A} = \{t^1_{\uparrow A}, t^2_{\uparrow A}, \cdots\}.$$

The *finish time* refers to the time point at which the finish event of an action occurs:

$$\downarrow A_T(t,i) : \mathbf{T} \times \mathbf{I}^+ \to \mathbf{B}$$

$$\forall t \in \mathbf{T} : \forall i \in \mathbf{I}^+ : [\downarrow A_T(t,i) \iff t = t^i_{\downarrow A}]$$

$$t^i_{\downarrow A} \in \mathbf{T}_{\downarrow A} = \{t^1_{\downarrow A}, t^2_{\downarrow A}, \cdots\}.$$

Similar to the point function definition, the value and time domain definitions of an interval function are alternative forms for describing the execution of the same action:

$$\forall t \in \mathbf{T} : \forall i \in \mathbf{I}^+ : [A_V(t,i) \iff A_T(t,i)].$$

The *execution time* of an action is the duration of the interval during which the action is executed.

In the following we state the two properties that are associated with the interval function:

duration property - the finish event of an action cannot precede the start event of the action:

$$\forall t, t' \in \mathbf{T} : \forall i \in \mathbf{I}^+ : [\downarrow A(t,i) \Rightarrow \uparrow A(t',i) \wedge t > t'].$$

If the start and finish events, occur at the same time point, implying a durationless action, then the action is modelled by a point function.

ordering property - two instances of the same action cannot be executed at the same time:

$$\forall t, t' \in \mathbf{T} : \forall i, j \in \mathbf{I}^+ : [\downarrow A(t,i) \wedge \uparrow A(t',j) \wedge i < j \Rightarrow t < t'].$$

Utility function

The specification of time uncertainties for the two functions defined above can be represented by means of a *utility function* $U(t,i)$, or value function [12]. We are concerned with the class of utility functions which are typically found in real-time safety-critical systems - *discrete* or *critical utility functions*. In these functions the utility, or usefulness, can only assume maximum and minimum values - hence $U(t,i)$ is a Boolean function.

The utility function $U_A(t,i)$ associated with the point function, which models the occurrence of an event, maps the timeline and the instance number of an event into the usefulness of its occurrence. The maximum usefulness in the occurrence of an event is obtained from the time interval established from the following two time attributes:

earliest occurring time (eot) - the first point in time at or after which an event can occur;

latest occurring time (lot) - the last point in time before which the event can occur.

An event either occurs during the interval of time established by these two time points or it is assumed not to have occurred. A utility function representing the usefulness of the occurrence of an event can be defined as follows:

$$U_E(t,i) : \mathbf{T} \times \mathbf{I}^+ \to \mathbf{B}$$

$$\forall t \in \mathbf{T} : \forall i \in \mathbf{I}^+ : [U_E(t,i) \iff (t^i_{E_{eot}} \leq t < t^i_{E_{lot}}$$

$$t^i_{E_{eot}} \in \mathbf{T}_{E_{eot}} = \{t^1_{E_{eot}}, t^2_{E_{eot}}, \cdots\}$$

$$t^i_{E_{lot}} \in \mathbf{T}_{E_{lot}} = \{t^1_{E_{lot}}, t^2_{E_{lot}}, \cdots\}.$$

The utility function $U_A(t,i)$ associated with the interval function, which models the execution of an action, maps the timeline and the instance number of an action into the usefulness of its execution. The maximum usefulness in the execution of an action is established by the two time intervals associated with the utility functions of the start and finish events of the action. The time attributes of the utility function associated with the execution of an action are the following:

earliest starting time (est)   -   the first point in time at or after which an action can start its execution;

latest starting time (lst)   -   the last point in time before which an action can start its execution;

earliest finishing time (eft)   -   the first point in time at or after which an action can finish its execution;

latest finishing time (lft)   -   the last point in time before which an action can finish its execution.

The time attributes *est* and *lft* are also referred to as "delay" and "deadline", respectively. These time attributes are expressed by step utility functions, respectively positive and negative step functions.

A utility function representing the usefulness in the execution of an action is defined as follows, in terms of the utility functions of the start and finish events of an action:

$$U_{\uparrow A}(t', i) : \mathbf{T} \times \mathbf{I}^+ \to \mathbf{B}$$

$$\forall t' \in \mathbf{T} : \forall i \in \mathbf{I}^+ : [U_{\uparrow A}(t', i) \iff (t^i_{A_{est}} \leq t' < t^i_{A_{lst}})]$$

$$t^i_{A_{est}} \in \mathbf{T}_{A_{est}} = \{t^1_{A_{est}}, t^2_{A_{est}}, \cdots\}$$

$$t^i_{A_{lst}} \in \mathbf{T}_{A_{lst}} = \{t^1_{A_{lst}}, t^2_{A_{lst}}, \cdots\};$$

$$U_{\downarrow A}(t'', i) : \mathbf{T} \times \mathbf{I}^+ \to \mathbf{B}$$

$$\forall t'' \in \mathbf{T} : \forall i \in \mathbf{I}^+ : [U_{\downarrow A}(t'', i) \iff (t^i_{A_{eft}} \leq t'' < t^i_{A_{lft}})]$$

$$t^i_{A_{eft}} \in \mathbf{T}_{A_{eft}} = \{t^1_{A_{eft}}, t^2_{A_{eft}}, \cdots\}$$

$$t^i_{A_{lft}} \in \mathbf{T}_{A_{lft}} = \{t^1_{A_{lft}}, t^2_{A_{lft}}, \cdots\};$$

$$U_A(t, i) : \mathbf{T} \times \mathbf{I}^+ \to \mathbf{B}$$

$$\forall t, t', t'' \in \mathbf{T} : \forall i \in \mathbf{I}^+ : [U_A(t, i) \iff U_{\uparrow A}(t', i) \wedge U_{\downarrow A}(t'', i)].$$

2.1.3   E/A model and time structures

One of the characteristics of the E/A model is that it can support both dense and discrete time structures. However, depending on the time struc-

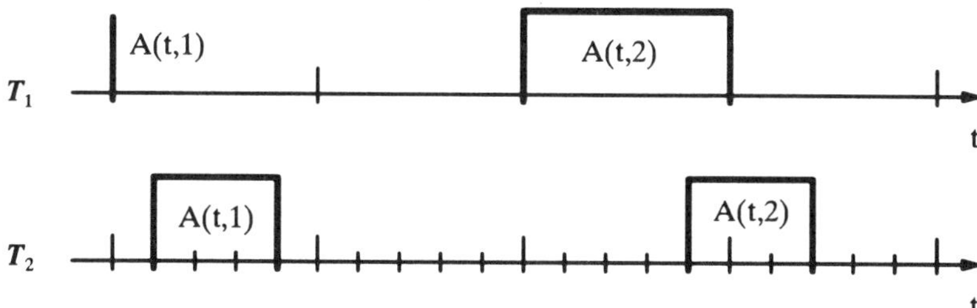

Figure 2. E/A model and the granularity of discrete time structures

ture there are some dissimilarities on how the occurrence of an event is observed. We say that an event is observed when a point on the timeline is associated with its occurrence.

In dense time structures the occurrence and observation of an event is always simultaneous, thus for distinct events we associate distinct time points. In discrete time structures, once an event occurs, it is not possible to observe an event itself, only its consequence(s) may be observed. Hence an event cannot be observed between any two time points on the timeline; the time point we associate with the observation of an event might not be coincident with the actual occurrence of the event.

Depending on the granularity of a discrete time structure (denoted by $\triangle(\mathbf{T})$), an action might be represented either by the point function or interval function. If we consider, for example, a time structure $\mathbf{T}_2$ which contains another time structure $\mathbf{T}_1$, as shown in Figure 2, then an action $A(t,1)$ when measured in $\mathbf{T}_2$ has a duration of less than $\triangle(\mathbf{T}_1)$, but when measured in $\mathbf{T}_1$ it becomes durationless. However, for another instance of the same action - $A(t,2)$, we can have a different situation, as shown in Figure 2. In summary, an action of fixed duration in a time structure $(\mathbf{T}_2)$, may assume different durations in another time structure $(\mathbf{T}_1)$ with a greater granularity - $\triangle(\mathbf{T}_1) > \triangle(\mathbf{T}_2)$, depending on the time points associated with the respective starting and finishing events of the action. (In order to maintain compatibility with the close/open interval being employed, we have opted to observe the occurrence of an event at the time point that precedes its actual occurrence.)

2.2 Predicate event/action notation (PEA notation)

The concern in the E/A model was to identify a set of primitive concepts and define a set of primitive functions that formalise these concepts. In order to express the primitive concepts of the E/A model concisely and to facilitate formal analysis of the system behaviour, we introduce the Predicate Event/Action notation (PEA notation). In the following, the primitive functions of the E/A model will be defined as primitive predicates of the PEA notation.

2.2.1   Primitive predicates of the PEA notation

Point predicate

For the value domain definition of the point predicate the notation "$E(i)$" is introduced to denote the ith instance of an event E. The point predicate is true iff there exists a time point t at which the point function $EV(t,i)$ holds true:

$$\forall i \in \mathbf{I}^+ : [E(i) \iff \exists t \in \mathbf{T} : E_V(t,i)].$$

For the time domain definition of the point predicate the notation "$E(i)@t$" is introduced to denote that the ith instance of an event E occurs at a time point t. The event predicate is true iff at the time point t the event E has occurred for the ith time:

$$\forall t \in \mathbf{T} : \forall i \in \mathbf{I}^+ : [E(i)@t \iff E_T(t,i)].$$

Interval predicate

For the value domain definition of the interval predicate the notation "$A(i)\langle \uparrow A, A_{\mathrm{INV}}, \downarrow A \rangle$" is introduced to denote the ith execution of action A. The start condition "$\uparrow A$" and finish condition "$\downarrow A$" are transition predicates, and the invariant condition "A_{INV}" is a state predicate. An action is executed between "$\uparrow A$" and "$\downarrow A$" while "A_{INV}" holds true. The conditions associated with an interval predicate are defined as follows:

$$\forall i \in \mathbf{I}^+ : [\uparrow A(i) \iff \exists t_{\uparrow A} \in \mathbf{T} : \uparrow A_V(t_{\uparrow A},i)],$$

$$\forall i \in \mathbf{I}^+ : [\uparrow A_{\mathrm{INV}}(i) \iff \exists t \in \mathbf{T} : A_V^{\mathrm{INV}}(t,i)], \text{ and}$$

$$\forall i \in \mathbf{I}^+ : [\downarrow A(i) \iff \exists t_{\downarrow A} \in \mathbf{T} : \downarrow A_V(t_{\downarrow A},i)].$$

The interval predicate is true iff for all time points between $t_{\uparrow A}$ and $t_{\downarrow A}$ the interval function $A_V(t,i)$ holds true. The interval predicate is defined in the value domain as follows:

$$\forall i \in \mathbf{I}^+ : [A(i)\langle \uparrow A, A_{\mathrm{INV}}, \downarrow A \rangle \iff \exists t_{\uparrow A}, t_{\downarrow A} \in \mathbf{T} : \forall t \in \mathbf{T} :$$

$$(t_{\uparrow A} \leq t < t_{\downarrow A} \iff A_V(t, i))].$$

For the time domain definition of the interval predicate the notation "$A(i)@\langle t_{\uparrow A}, t_{\downarrow A}\rangle$" is introduced to denote the ith execution of an action A between the time points $t_{\uparrow A}$ and $t_{\downarrow A}$. These time points represent the times at which the events associated with action A have occurred: the start event at time $t_{\uparrow A}$ and the finish event at time $t_{\downarrow A}$. The two events are respectively defined as follows, in terms of the point function:

$$\forall t_{\uparrow A} \in \mathbf{T} : \forall i \in \mathbf{I}^+ : [\uparrow A(i)@t_{\uparrow A} \iff \uparrow A_T(t_{\uparrow A}, i)], \text{ and}$$

$$\forall t_{\downarrow A} \in \mathbf{T} : \forall i \in \mathbf{I}^+ : [\downarrow A(i)@t_{\downarrow A} \iff \downarrow A_T(t_{\downarrow A}, i)].$$

The interval predicate is true iff for all time points between $t_{\uparrow A}$ and $t_{\downarrow A}$ the interval function $A_T(t, i)$ holds true. The interval predicate is defined in the time domain as follows:

$$\forall t_{\uparrow A}, t_{\downarrow A} \in \mathbf{T} : \forall i \in \mathbf{I}^+ : [A(i)@\langle t_{\uparrow A}, t_{\downarrow A}\rangle \iff \forall t \in \mathbf{T} :$$

$$(t_{\uparrow A} \leq t < t_{\downarrow A} \iff A_T(t, i))].$$

Utility predicate

In the following we define the utility predicates that will be employed in the reasoning of time uncertainties.

The utility predicate "$U_E(i)@\langle t_{eot}, t_{lot}\rangle$" is introduced to denote the time uncertainty associated with the ith occurrence of event E. The utility predicate is true iff for all time points between t_{eot} and t_{lot} the utility function $U_E(t, i)$ holds true. The utility predicate representing the usefulness in the occurrence of an event is represented as:

$$\forall t_{eot}, t_{lot} \in \mathbf{T} : \forall i \in \mathbf{I}^+ : [U_E(i)@\langle t_{eot}, t_{lot}\rangle \iff \forall t \in \mathbf{T} :$$

$$(t_{eot} \leq t < t_{lot} \iff U_E(t, i))].$$

For the utility predicate of an action A, the notation "$U_A(i)@\langle t_{\uparrow A_{est}}, t_{\uparrow A_{lst}}, t_{\downarrow A_{eft}}, t_{\downarrow A_{lft}}\rangle$, is introduced to denote the time uncertainty associated with the ith execution of the action. The utility predicate is true iff for all time points between $t_{\uparrow A_{est}}$, and $t_{\uparrow A_{lst}}$, the utility function $U_{\uparrow A}(t', i)$ holds true, and for all time points between $t_{\downarrow A_{eft}}$, and $t_{\downarrow A_{lft}}$, the utility function $U_{\downarrow A}(t'', i)$ holds true. The utility predicate representing the usefulness in the execution of an action is represented as:

$$\forall t_{\uparrow A_{est}}, t_{\uparrow A_{1st}}, t_{\downarrow A_{eft}}, t_{\downarrow A_{lft}} \in \mathbf{T} :$$

$$\forall i \in \mathbf{I}^+ : [U_A(i)@\langle t_{\uparrow A_{est}}, t_{\uparrow A_{1st}}, t_{\downarrow A_{eft}}, t_{\downarrow A_{lft}} \rangle \Longleftrightarrow$$

$$\forall t' \in \mathbf{T} : (t_{\uparrow A_{est}} \leq t' < t_{\uparrow A_{lst}} \Longleftrightarrow U_{\uparrow A}(t', i)) \wedge$$

$$\forall t'' \in \mathbf{T} : (t_{\downarrow A_{eft}} \leq t'' < t_{\downarrow A_{lft}} \Longleftrightarrow U_{\downarrow A}(t'', i))].$$

2.2.2   Operators of the PEA notation

In order to compose point and interval predicates of the PEA notation, a set of logical operators are defined in both value and time domains. In the following, we define some of the PEA notation operators in term of the functions of the E/A model.

From two standard logical operators, negation ($\neg$) and conjunction ($\wedge$), other logical operators can be defined, such as disjunction ($\vee$), implication ($\Rightarrow$) and equivalence ($\Longleftrightarrow$). As an example, we define, in the value domain, the conjunction of two interval predicates as follows:

$$\forall i, j \in \mathbf{I}^+ : A(i)\langle \uparrow A, A_{\mathrm{INV}}, \downarrow A \rangle \wedge B(j)\langle \uparrow B, B_{\mathrm{INV}}, \downarrow B \rangle \overset{\mathrm{def}}{=}$$

$$\exists t \in \mathbf{T} : A_V(t, i) \wedge B_V(t, j).$$

Four additional logical operators were defined over the interval predicates of the PEA notation: choice ($+$), meet ($\prec$), overlap ($\|$) and disjoint (∇). As an example, we define, in both value and time domains, the meet and overlap operators.

The meet operator ($\prec$):

$$\forall i, j \in \mathbf{I}^+ : A(i)\langle \uparrow A, A_{\mathrm{INV}}, \downarrow A \rangle \prec B(j)\langle \uparrow B, B_{\mathrm{INV}}, \downarrow B \rangle \overset{\mathrm{def}}{=}$$

$$A(i)\langle \uparrow A, A_{\mathrm{INV}}, \downarrow A \rangle \wedge B(j)\langle \uparrow B, B_{\mathrm{INV}}, \downarrow B \rangle \wedge \downarrow A(i) \Longleftrightarrow \uparrow B(j);$$

$$\forall t_{\uparrow A}, t_{\downarrow A}, t_{\uparrow B}, t_{\downarrow B} \in \mathbf{T} : \forall i, j \in \mathbf{I}^+ : A(i)@\langle t_{\uparrow A}, t_{\downarrow A} \rangle \prec B(j)@\langle t_{\uparrow B}, t_{\downarrow B} \rangle \overset{\mathrm{def}}{=}$$

$$A(i)@\langle t_{\uparrow A}, t_{\downarrow A} \rangle \wedge B(j)@\langle t_{\uparrow B}, t_{\downarrow B} \rangle \wedge (t_{\downarrow A} = t_{\uparrow B}).$$

The overlap operator ($\|$):

$$\forall i,j \in \mathbf{I}^+ : A(i)\langle\uparrow A, A_{\mathrm{INV}}, \downarrow A\rangle \| B(j)\langle\uparrow B, B_{\mathrm{INV}}, \downarrow B\rangle \stackrel{\mathrm{def}}{=}$$

$$A(i)\langle\uparrow A, A_{\mathrm{INV}}, \downarrow A\rangle \wedge B(j)\langle\uparrow B, B_{\mathrm{INV}}, \downarrow B\rangle \wedge$$

$$\uparrow A(i) \Rightarrow B(j)\langle\uparrow B, B_{\mathrm{INV}}, \downarrow B\rangle \vee \uparrow B(j) \Rightarrow A(i)\langle\uparrow A, A_{\mathrm{INV}}, \downarrow A\rangle;$$

$$\forall t_{\uparrow A}, t_{\downarrow A}, t_{\uparrow B}, t_{\downarrow B} \in \mathbf{T} : \forall i,j \in \mathbf{I}^+ : A(i)@\langle t_{\uparrow A}, t_{\downarrow A}\rangle \| B(j)@\langle t_{\uparrow B}, t_{\downarrow B}\rangle \stackrel{\mathrm{def}}{=}$$

$$(A(i)@\langle t_{\uparrow A}, t_{\downarrow A}\rangle \wedge B(j)@\langle t_{\uparrow B}, t_{\downarrow B}\rangle) \wedge ((t_{\uparrow B} \leq t_{\uparrow A} < t_{\downarrow B}) \vee$$

$$(t_{\uparrow A} \leq t_{\uparrow B} < t_{\downarrow A})).$$

Apart from the above two sets of operators, we define another special operator that is only applicable to the time domain; the *duration operator* (or δ-*operator*) measures the time distance between two time points (associated with the occurrence of events). The duration between the occurrence of the events $E(i)@t_E$ and $F(i)@t_F$ is given by the duration operator, as follows:

$$\delta_{(t,t)} : \mathbf{T} \times \mathbf{T} \to \mathbf{T}$$

$$\delta_{(t_{\uparrow E}, t_{\uparrow F})} = t_{\uparrow F} - t_{\uparrow E}.$$

The minimum and maximum durations between the occurrence of two events is obtained from their respective utility predicates, by using the following two variations of the duration operator:

$$\delta_{(t,t)}^{\min} : \mathbf{T} \times \mathbf{T} \to \mathbf{T} \qquad\qquad \delta_{(t,t)}^{\max} : \mathbf{T} \times \mathbf{T} \to \mathbf{T}$$

$$\delta_{(t_{\uparrow E}, t_{\uparrow F})}^{\min} = t_{\uparrow F_{eot}} - t_{\uparrow E_{lot}} \qquad\qquad \delta_{(t_{\uparrow E}, t_{\uparrow F})}^{\max} = t_{\uparrow F_{lot}} - t_{\uparrow E_{eot}}.$$

If the two time points of the duration operator refer to the time of occurrence of the start and finish events of an action, then the notation $\delta_{A(i)}$ can be used instead of $\delta_{(t_{\uparrow A}^i, t_{\downarrow A}^i)}$.

2.2.3 The PEA notation and other techniques

Of the many possible formal techniques that could be employed in the context of the proposed approach, the PEA notation is compared with three similar techniques: RTL [4], TRIO [13] and VVSL [14]. RTL captures temporal properties of a system in terms of events and actions. To permit the mechanical reasoning of RTL formulae the expressiveness of the logic was restricted. Specifications in RTL are built using an occurrence function, which relate the occurrence of an event to a time point, this makes it difficult to separate the analysis of the value and time domains. TRIO is a first-order temporal logic language for specifying and verifying timing requirements; its proof theory and its executability can be mathematically defined. The language provides operators that allow the truth or falsity of a proposition at particular time instants. TRIO can accommodate either dense or discrete time structures. However, TRIO does not provide mechanisms for modularizing complex specifications, and it is hard to read and understand. VVSL defines operations in terms of temporal predicates over state variables, in addition to the pre-condition and post-condition of VDM, VVSL introduces an *inter-condition* which imposes constraints during execution (i.e. acts as a state-invariant). However, unlike the PEA notation, VVSL does not permit quantitative timing analysis.

3   Extract of a case study: controlling the temperature in a nuclear reactor

To clarify some concepts introduced so far, an extract of a case study based on a simplified nuclear reactor control system was selected as an example [15]. Specifically, the example will serve to illustrate how the concepts of the E/A model can be incorporated within different classes of formalisms in order to describe system behaviour.

The example involves a system used to control the temperature of a nuclear reactor, the rods of the reactor have to be moved down when the temperature reaches the pre-defined threshold (5000K). The activity of moving down the rods takes 20 time units, and the start of consecutive movements of the rods should be at least 30 time units apart. To simplify the concepts, we make the (strong) assumption that the movement of the rods is not constrained by any physical limitation.

The case study from the PEA notation perspective

In a very simplified form, the nuclear plant can be defined, in the PEA notation, as the parallel composition of the actions describing the physical process and safety controller. ($A(i)\langle\rangle$ and $A(i)@\langle\rangle$ are abbreviations for the notation previously introduced.)

$$\forall i \in \mathbf{I}^+ : Nuclear\_Plant(i)\langle\rangle \Longleftrightarrow$$

$$Physical\_Process(i)\langle\rangle \| Safety\_Controller(i)\langle\rangle.$$

The physical process is defined by the sequential composition of actions representing the down movement of the rods ($PDMovRods(i)\langle\rangle$) and no movement of the rods ($PNDMovRods(i)\langle\rangle$).

$$\forall i \in \mathbf{I}^+ : Physical\_Process(i)\langle\rangle \iff$$

$$PNDMovRods(i)\langle\rangle \prec PDMovRods(i)\langle\rangle.$$

In order to capture the timing requirements imposed on the actions the δ-operator can be employed to specify that the execution of $PDMovRods(i)\langle\rangle$ should take exactly 20 units of time, and consecutive executions should be at least 30 units of time apart:

$$\forall i \in \mathbf{I}^+ : \delta_{PDMovRods(i)} = 20.$$

$$\forall i \in \mathbf{I}^+ : \delta^{\min}_{(t^i_{\uparrow PDMovRods}, t^{i+1}_{\uparrow PDMovRods})} = 30.$$

The safety controller is defined by sequential composition of actions representing the down movement of the rods $(CoDMovRods(i)\langle\rangle)$, no movement $(CoNDMovRods(i)\langle\rangle)$, and a wait action on the down movement of the rods $(CoWMovRods(i)\langle\rangle)$.

$$\forall i \in \mathbf{I}^+ : Safety\_Controller(i)\langle\rangle \iff$$

$$CoNDMovRods(i)\langle\rangle \prec CoDMovRods(i)\langle\rangle \prec CoWMovRods(i)\langle\rangle.$$

In Sections 4.1 and 4.2, we show how the PEA notation specifications can be used as templates for specifying the behaviour in a property-oriented formalism (THL) and an operational formalism (ER nets).

4 E/A model and other formalisms

In this section, we illustrate how the primitive concepts of the E/A model can be incorporated into other formalisms. The general approach adopted is to define the primitive functions in terms of the primitives of a formalism, and then impose restrictions over these primitives to capture the basic properties of the point and interval functions. This approach is applied to the classes of formalisms identified in this paper: property-oriented and operational.

Property-oriented formalisms specify behaviour in terms of the properties that are exhibited by a system. A property-oriented specification is axiomatic, and hence has a conjunctive nature which allows properties to be added or removed (during analysis) without the need to reconstruct the full specification. A property-oriented specification should only state the necessary constraints on system behaviour (no explicit interrelationships are imposed between parts of the specification) to minimise the restrictions imposed on possible implementations. Temporal Logic [16] and Timed History Logic (THL) [17] are examples of property-oriented formalisms. Operational formalisms specify behaviour by constructing an executable model of the system in terms of mathematical structures such as tuples, relations, functions, sets and sequences. In our approach, we are only concerned with those operational formalisms that explicitly specify non-determinism and concurrency. Statecharts [5] and Environment/Relationship nets (ER nets) [18] are examples of such operational formalisms.

4.1 Property-oriented formalisms

In the following we present how the primitive functions of the E/A model can be incorporated into THL [17]. THL is a logical formalism, based on the notion of system histories. A history H of a system is a function of the form $H : \mathbf{T} \to \Gamma$; where $\mathbf{T}$ is a closed time interval representing the operational lifetime of the system and Γ the state space of the system (i.e. the cross product of the set of values for all the variables in a state vector SV). The set of all "possible" histories of a system are defined as the universal history set $\Gamma\mathbf{H}$. For a history H the sequence of values taken by a variable v_i of SV is denoted by the function $H \cdot v_i : \mathbf{T} \to V_{v_i}$.

Specifications are expressed by restricting the set of histories by two sorts of relations. *Invariant relations* are formulated as predicates built using free value variables for each v_i of SV. A history satisfies an invariant relation if and only if the predicate is satisfied at every time point within $\mathbf{T}$. *History relations* are formulated as predicates built using two free time variables T_0, T_1 and free function variables for each v_i of SV (T_0 and T_1 should be interpreted as being universally quantified over $\mathbf{T}$). A history satisfies a history relation if and only if the predicate holds for every pair of points $T_0 < T_1$ within $\mathbf{T}$.

The THL description of an E/A model of a system is obtained by extending the state space Γ, in terms of the primitive functions. That is, for a system with q point functions and r interval functions the state vector is extended by the following variables:

$$\langle E_1(t), \cdots, E_q(t), \uparrow A_1(t), \cdots, \uparrow A_r(t), \downarrow A_1(t), \cdots, \downarrow A_r(t), A_1(t), \cdots, A_r(t)\rangle.$$

The above functions are of the form $E(t) : \mathbf{T} \to \mathbf{B} \times \mathbf{I}^+$. However, to be consistent with the E/A model the following convention is adopted to introduce the parameter "i":

$$\forall H \in \Gamma\mathbf{H} : \forall t \in \mathbf{T} : \forall i \in \mathbf{I}^+ : [H \cdot E(t, i) \equiv H \cdot E(t) = (true, i)].$$

Axioms are then introduced into the THL description, to ensure that the *uniqueness* and *ordering* properties of point functions and the *duration* and *ordering* properties of interval functions are satisfied for all well-defined histories. For example, we say that a history H is well-defined for a point function E if and only if:

$$\forall t, t' \in \mathbf{T} : \forall i \in \mathbf{I}^+ : [H \cdot E(t, i) \wedge H \cdot E(t', i) \Rightarrow t = t'].$$

The bar operator "$\mid^C$" of the E/A model is defined in THL as the following history relation:

$$\forall t \in \mathbf{T} : \mid^C (sp(t)) \Longleftrightarrow$$

$$\forall t_1 \in \mathbf{T} : (T_1 = t \wedge sp(T_1) \wedge \exists t_1 \in \mathbf{T} : \forall t_2 \in \mathbf{T} : t_1 < t_2 < T_1 \Rightarrow \neg sp(t_2)).$$

The case study from the property-oriented formalism perspective

The behaviour of the nuclear plant is specified in THL by imposing invariant and history relations over the start and finish conditions of the actions of the physical process and safety controller.

Physical process

Pr1.   The initial condition of the physical process is that $PNDMovRods$ holds true.

$$\forall T_1 \in \mathbf{T} : PNDMovRods(1)(T_1) \Longleftrightarrow (T_1 = 0).$$

Pr2.   The rods must start to move down if and only if the temperature ($PTemp$) is above the threshold (5000K) and 30 time units have elapsed since the previous time the rods started to move down, the execution of the action should take exactly 20 time units.

$$\forall T_1 \in \mathbf{T} : \uparrow PDMovRods(1)(T_1) \Longleftrightarrow$$

$$(PTemp > 5000)(T_1) \wedge \forall t \in \mathbf{T} : t < T_1 \Rightarrow \neg(PTemp > 5000)(t).$$

$$\forall T_1 \in \mathbf{T} : \forall i \in \mathbf{I}^+ : i > 1 \Rightarrow \uparrow PDMovRods(i)(T_1) \Longleftrightarrow$$

$$(PTemp > 5000)(T_1) \wedge \exists t_1 \in \mathbf{T} : (t_1 \leq T_1 - 30 \wedge$$

$$\uparrow PDMovRods(i-1)(t_1) \wedge \forall t \in \mathbf{T} : t_1 + 30 \leq t < T_1 \Rightarrow$$

$$\neg(PTemp > 5000)(t)).$$

$$\forall T_1 \in \mathbf{T} : \forall i \in \mathbf{I}^+ : \downarrow PDMovRods(i)(T_1 + 20) \Longleftrightarrow$$

$$PDMovRods(i)(T_1).$$

Pr3.   *PNDMovRods* immediately follows the action *PDMovRods*, as specified by the sequential composition of the actions in the PEA notation description of *Physical_Process*.

$$\forall i \in \mathbf{I}^+ : \uparrow PNDMovRods(i+1) \Longleftrightarrow \downarrow PDMovRods(i).$$

$$\forall i \in \mathbf{I}^+ : \downarrow PNDMovRods(i) \Longleftrightarrow \uparrow PDMovRods(i).$$

Safety controller

In this example, we assume that provided the reactor temperature (*PTemp*) is within its anticipated range then it is equal to the thermometer reading (*CiTemp*).

Cr1.   The initial condition of the safety controller is that *CoNDMovRods* holds true.

$$\forall T_1 \in \mathbf{T} : CoNDMovRods(1)(T_1) \Longleftrightarrow (T_1 = 0).$$

Cr2.   The safety controller must start to move the rods down at the instant the temperature (*CiTemp*) rises above the threshold (5000K) and 30 time units have elapsed since the previous time the rods started to move down, the execution of the action *CoDMovRods* should take exactly 20 time units.

$$\forall T_1 \in \mathbf{T} : \uparrow CoDMovRods(1)(T_1) \Longleftrightarrow$$

$$(CiTemp > 5000)(T_1) \land \forall t \in \mathbf{T} : t < T_1 \Rightarrow \neg(CiTemp > 5000)(t).$$

$$\forall T_1 \in \mathbf{T} : \forall i \in \mathbf{I}^+ : i > 1 \Rightarrow \uparrow CoDMovRods(i)(T_1) \Longleftrightarrow$$

$$(CiTemp > 5000)(T_1) \land \exists t_1 \in \mathbf{T} : (t_1 \leq T_1 - 30 \land$$

$$\uparrow CoDMovRods(i-1)(t_1) \land \forall t \in \mathbf{T} : t_1 + 30 \leq t < T_1 \Rightarrow$$

$$\neg(CiTemp > 5000)(t)).$$

$$\forall T_1 \in \mathbf{T} : \forall i \in \mathbf{I}^+ : \downarrow CoDMovRods(i)(T_1 + 20) \Longleftrightarrow$$

$$CoDMovRods(i)(T_1).$$

Cr3. *CoWMovRods* immediately follows action *CoDMovRods*, and is of duration exactly 10 time units.

$$\forall i \in \mathbf{I}^+ : \uparrow CoWMovRods(i) \Longleftrightarrow \downarrow CoDMovRods(i).$$

$$\forall T_1 \in \mathbf{T} : \forall i \in \mathbf{I}^+ : \downarrow CoWMovRods(i)(T_1 + 10) \Longleftrightarrow$$

$$\uparrow CoWMovRods(i)(T_1).$$

Cr4. *CoNDMovRods* must immediately follow *CoWMovRods* and precede *CoDMovRods*, as specified by the sequential composition of the actions in the PEA notation description of *Safety_Controller*.

$$\forall i \in \mathbf{I}^+ : \uparrow CoNDMovRods(i+1) \Longleftrightarrow \downarrow CoWMovRods(i).$$

$$\forall i \in \mathbf{I}^+ : \downarrow CoNDMovRods(i) \Longleftrightarrow \uparrow CoDMovRods(i).$$

4.2 Operational formalisms

In the following we present how the primitive functions of the E/A model can be incorporated into ER nets [18]. ER nets are high-level Petri nets where tokens carry information (functions associating values to variables), and transitions are augmented with predicates (describing which input tokens can participate in a firing and which possible tokens are produced by the firing). The timing constraints are expressed as relational expressions (predicates) which must be satisfied for a transition to become enabled.

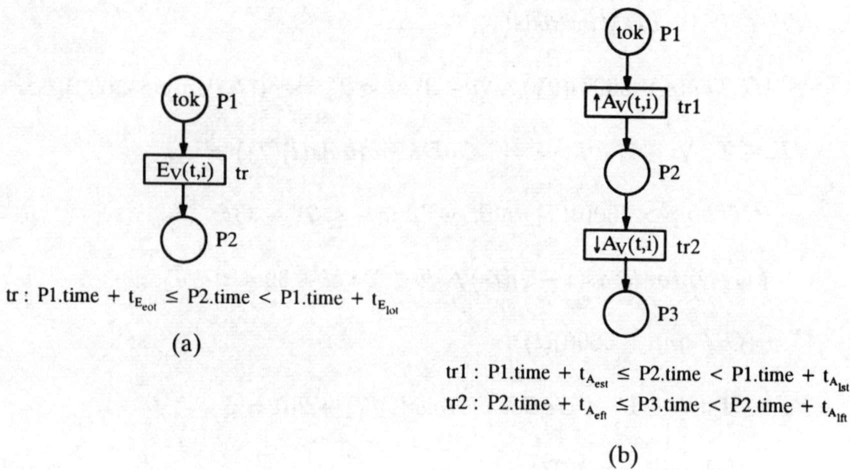

Figure 3. The ER nets representation of the E/A model: (a) point function and (b) interval function

The representation of the E/A model primitive functions in terms of ER nets is straight forward because of the similarity between the firing rules of a transition and the properties associated with the occurrence of events. The timing properties defined for both point and interval functions of the E/A model are captured by the timing predicates to be associated with the transitions.

The point function in the E/A model corresponds in ER nets to the firing of a transition. Figure 3(a) shows the representation of the point function in terms of ER nets. The label $E_V(t,i)$ represents the predicate that defines the occurrence of an event. When transition tr fires the token *tok* is removed from predicate $P1$ and placed in $P2$ with a timestamp $P2.time(= t^i_E)$ corresponding to the time at which the transition fired. The instance number of an event is captured in the predicate of tr, by increasing the value of indices i. Time uncertainties can also be modelled by associating a utility function with the occurrence of the event, once transition tr is enabled, it is allowed to fire only within the time interval established by the timing relation associated with the transition.

The representation of the E/A model interval function follows directly from the point function representation. The labels $\uparrow A_V(t,i)$ and $\downarrow A_V(t,i)$ on transitions $tr1$ and $tr2$ represent, respectively, the start and finish events that are associated with the execution of an action, and place $P2$ denotes the execution of the action. The invariant condition associated with the in-

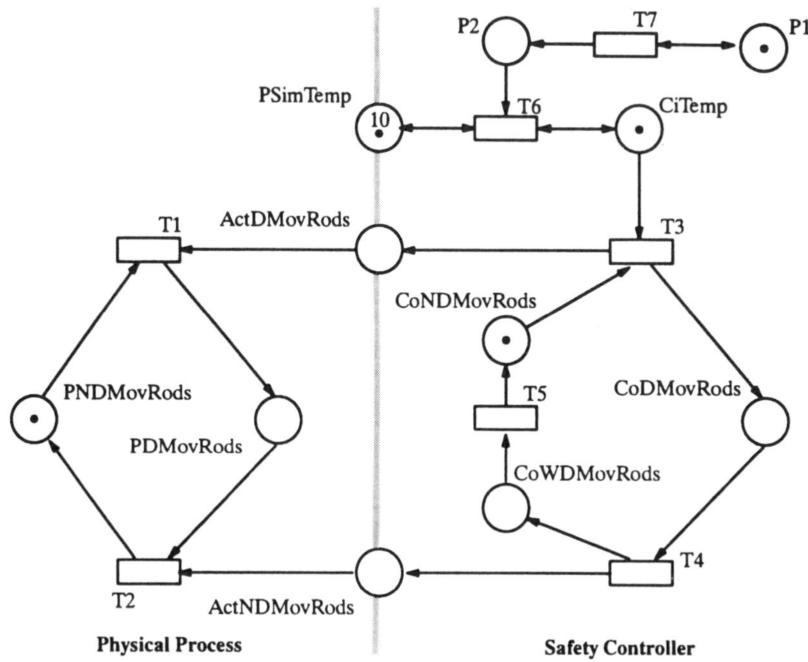

Legend of the relevant transitions:

$T3 = \{\langle\langle CiTemp, CoNDMovRods\rangle, \langle CoDMovRods, ActDMovRods\rangle\rangle \mid CiTemp.value > 5000\}$

$T4 = \{\langle\langle CoDMovRods, ActNDMovRods\rangle, CoWDMovRods\rangle \mid$
$\qquad CoWDMovRods.time = CoDMovRods.time + 20\}$

$T5 = \{\langle CoWDMovRods, CoNDMovRods\rangle \mid CoNDMovRods.time = CoWDMovRods.time + 10\}$

$T7 = \{\langle P1, \langle P1, P2\rangle\rangle \mid P2.time = P1.time + 15\}$

Figure 4. The ER net model of the nuclear reactor

terval function $(A_V{}^{INV}(t, i))$ is part of the predicate of transition $tr1$. The violation of $A_V{}^{INV}(t, i)$ while, for example, an action is being executed, can be represented by a choice on $P2$ between transition $tr2$ and another transition which negates the invariant condition. Figure 3(b) shows the representation of the interval function in terms of ER nets. Time uncertainties can also be modelled by associating a utility function with the occurrence of each event that defines the execution of an action.

The case study from the operational formalism perspective

In order to exemplify the utilization of the E/A model primitive functions as modelling concepts for an operational formalism, in the following, we discuss the ER net model of the movement of the rods to control the temperature of a nuclear reactor. The ER net model is shown in Figure 4 and includes models of the physical process and the safety controller. The name and the role of the places of the ER net correspond to the actions specified by the property-oriented formalism, presented in the previous section. The relations associated with the transitions of the ER net are defined in Figure 4, and correspond to the conditions of the actions defined in THL. In the ER net model, the place *CiTemp* contains the last measurement of the reactor temperature which is periodically updated from the simulated values contained in *PSimTemp*. The subnet containing *PDMovRods* and *PNDMovRods* is obtained from the PEA notation formula that defines the sequence of actions that are associated with the rods; the subnet containing *CoDMovRods*, *CoWDMovRods* and *CoNDMovRods* is obtained from the PEA notation formula that defines the sequence of actions that have to be performed by the safety controller in order to meet the requirements imposed on the movement of the rods depending on the temperature of the reactor; the places *ActDMovRods* and *ActNDMovRods* represent the outputs of the (logical) actuator that moves the rods. The predicates associated with the transitions $T3$, $T4$ and $T5$ are derived from the start and finish conditions defined for the actions to be executed by the safety controller.

5   Conclusions and future work

This paper presents the E/A model, as a basic model for describing the behaviour of real-time safety-critical systems, which consists of primitive concepts such as events, actions and states. The concepts of the E/A model are related to the flow of time by introducing two primitive functions: *point function* and *interval function*. A key feature of the E/A model is its flexibility: it can be incorporated into a wide range of formalisms which are based on either dense or discrete time structures, and supports analysis in either the time domain or value domain.

For the E/A model a formal syntax and semantics are defined which precisely characterise the concepts of events, states and actions. The PEA notation is introduced to facilitate the formal analysis of the primitive concepts of the E/A model. Currently, the PEA notation provides a formal syntax for the primitive concepts of the E/A model and defines a set of operators for the composition of these concepts. The semantics of the PEA notation are described in terms of the E/A model. The flexibility of the E/A model is demonstrated by introducing the primitive functions, into a property-oriented formalism (THL) and an operational formalism (ER nets). The support, for modelling system behaviour, provided by the primitive concepts of the E/A model was illustrated by performing the analysis of a simple example from the perspectives of the PEA notation, THL and ER nets.

An important issue, still to be addressed, is the provision of a proof theory for the PEA notation. One approach is to incorporate the concepts of the E/A model and syntax of the PEA notation into a formalism with a well-defined proof theory. This has the following potential advantages: an appropriate base formalism would enhance the formal foundations of the PEA notation by providing a strong basis for a proof-theory and other support such as mechanical proof checkers; the PEA notation would enhance the applicability of the base formalism, during the requirements analysis of process control systems, by providing more appropriate concepts and operators. Currently, three formalisms are under investigation: Extended Duration Calculus [1], TLA+ [19] and TRIO [13].

Acknowledgements

The authors would like to acknowledge the anonymous reviewer for the helpful comments and suggestions, and the financial support of British Aerospace (DCSC) and SERC (UK) SCHEMA project.

Bibliography

1. Chaochen, Z., Ravn, A.P. and Hansen, K.M. (1993). An extended duration calculus for hybrid real-time systems. *Hybrid Systems*, Lecture Notes in Computer Science, *736*, Editors: R.L. Grossman, A. Nerode, A.P. Ravn and H. Rischel, Springer-Verlag, 36-59.

2. Liu, Z., Nordahl, J. and Sorensen, E.V. (1993). Composition and refinement of probabilistic real-time systems. *Proceedings of the International Conference on Computer Safety, Reliability and Security (SAFECOMP'93)*, Editor: J. Górski, Springer-Verlag, pp. 31-40.

3. Hansson, H. and Jonson, B. (1990). A calculus for communicating systems with time and probabilities. *Proceedings of the 11th Real-Time Systems Symposium*, Lake Buena Vista, FL, pp. 278-287.

4. Jahanian, F. and Mok, A. (1986). Safety analysis of timing properties in real-time systems. *IEEE Transactions on Software Engineering, SE-12(9)*, 890-904.

5. Harel, D. (1987). Statecharts: A visual formalism for complex systems. *Science of Computer Programming, 8*, 231-274.

6. Saeed, A., de Lemos, R. and Anderson, T. (1993). Robust requirements specifications for safety-critical systems. *Proceedings of the 12th International Conference on Computer Safety, Reliability and Security (SAFECOMP'93)*, Editor: J. Górski, Springer-Verlag, pp. 219-229.

7. de Lemos, R., Saeed, A. and Anderson, T. (1992). Analysis of timeliness requirements in safety-critical systems. *Proceedings of the Symposium in Formal Techniques in Real-Time and Fault-Tolerant Systems*, Editor: J. Vytopil, *Lecture Notes in Computer Science*, *571*, Springer-Verlag, pp. 171-192.

8. Anderson, T., de Lemos, R., Fitzgerald, J.S. and Saeed, A. (1993). On formal support for industrial-scale requirements analysis. *Hybrid Systems, Lecture Notes in Computer Science*, *736*, Editors: R.L. Grossman, A. Nerode, A.P. Ravn and H. Rischel, Springer-Verlag, 426-451.

9. Heninger, K.L. (1980). Specifying software requirements for complex systems: New techniques and their applications. *IEEE Transactions on Software Engineering*, *SE-6(1)*, 2-13.

10. de Lemos, R., Saeed, A. and Anderson, T. (1991). Value inconsistencies due to time uncertainties. *Proceedings of the 10th IFAC Workshop on Distributed Computer Control Systems*, Editors: H. Kopetz and M.G. Rodd, pp. 1-6.

11. Dasarathy, B. (1985). Timing constraints of real-time systems: Constructs for expressing them, methods of validating them. *IEEE Transactions on Software Engineering*, *SE-11(1)*, 80-86.

12. Jensen, E., Locke, D. and Tokuda, H. (1985). A time-driven scheduling model for real-time operating systems. *Proceedings of the 6th Real-Time Systems Symposium*, pp. 112-122.

13. Ghezzi, C., Mandrioli, D. and Morzenti, A. (1990). TRIO: A logic language for executable specifications of real-time systems. *Journal of Systems and Software*, *12*, 107-123.

14. Middleburg, C.A. (1989). VVSL: A language for structured VDM specifications. *Formal Aspects of Computing*, *1*, 115-135.

15. Saeed, A., de Lemos, R. and Anderson, T. (1993). Formal techniques for requirements analysis for safe reactor control. *The Nuclear Engineer*, *34(4)*, 108-115.

16. Manna, Z. and Pnueli, A. (1992). *The temporal logic of reactive and concurrent systems*, Springer-Verlag, New York, USA.

17. Saeed, A., Anderson, T. and Koutny, M. (1990). A formal model for safety-critical computing systems. *Proceedings of the International Conference on Computer Safety, Reliability and Security (SAFE-COMP'90)*, Editor: B.K. Daniels, pp. 1-6.

18. Ghezzi, C., Mandrioli, D., Morasca, S. and Pezzè, M. (1991). A unified high-level petri net formalism for time-critical systems. *IEEE Transactions on Software Engineering, SE-17(2)*, 160-172.

19. Lamport, L. (1993). Hybrid Systems in TLA+. *Hybrid Systems, Lecture Notes in Computer Science, 736*, Editors: R.L. Grossman, A. Nerode, A.P. Ravn and H. Rischel, Springer-Verlag, 77-102.

Proving authentication protocols – what do authentication protocols prove?

Dieter Gollmann

Department of Computer Science, Royal Holloway, University of London

Abstract An attempt is presented at showing how the BAN logic could have been used to detect flaws in authentication mechanisms discussed during the drafting of the international standard ISO/IEC 9798-3. Requirements on authentication mechanisms are proposed which could serve as the basis for formal proofs and which would have exposed a flawed mechanism in a draft of the standard.

1 Introduction

International standards for security are being developed by several committees within the International Organization for Standardization (ISO). Among these committees, ISO/IEC JTC1/SC27 works on the standardization of authentication mechanisms. A very recent result of these activities is the standard ISO/IEC 9798-3 dealing with *entity authentication using a public key algorithm* [9]. This paper will look at the history of this standard. During its emergence, mechanisms were improved, hopefully, found to be flawed, improved again, until this process terminated in a voting procedure. In [11] the question was put forward how formal methods could have been employed in this process and, more specifically, whether a formal analysis would have detected a flaw in a particular 'improved' mechanism.

We will show that an attempted proof using the BAN logic [1] would indeed have indicated a potential weakness of that mechanism. To do so, we do not have to introduce new operators to the BAN logic but rather answer the question

What Do Authentication Protocols Prove?

for challenge/response (handshake) authentication mechanisms. To pursue this question, we extract from the introduction to ISO/IEC 9798-1 the following definitions.

Entity authentication mechanisms allow the verification, of an entity's claimed entity, by another entity ... An impersonator may replay, at a later date, a valid authentication exchange (this is a form of masquerade). To prevent such a replay, a time variant parameter, such as a time stamp, a sequence number, or a challenge may be used ...

In ISO/IEC 9798-3, the authentication mechanisms should meet the following requirements.

In the authentication mechanisms specified in this part of ISO/IEC 9798 an entity to be authenticated corroborates its identity by demonstrating its knowledge of its secret signature key. This is achieved by the entity using its signature to sign specific data. The signature can be verified by anyone using the entity's public verification key.

2   The BAN logic

The BAN logic was proposed as a *Logic of Authentication*. The authors express the purpose of an authentication mechanism in terms of the beliefs held by the participants after having completed an authentication. They refrain, however, from giving a precise definition of the beliefs an authentication mechanism should generate. For good reasons, they then concentrate almost entirely on authentic key exchange. In [1], an analysis of nine mechanisms is presented. In eight of the nine examples, the purpose of the mechanism is key exchange. The noteable exception is the three-message authentication mechanism from CCITT X.509. It is very much a mechanism of the type discussed in ISO/IEC 9798-3, using time stamps, nonces (challenges), and public key signatures. Two parties, A and B, attempt a mutual authentication.

- In the first message, A sends a signed challenge to B,

- in the second message, B responds to A's challenge and includes its own signed challenge,

- in the third message, A replies to B's challenge.

There is no direct link between the first and third message so B cannot be sure that A's final response and A's initiating message are part of the same authentication attempt. Indeed, an attacker can use this feature to impersonate A. In the language of the BAN logic, we lack a proof that, at the end of the message exchange, A *believes* in the first message.

Maybe, it should not come as a surprise that the BAN logic had its initial success with an authentication mechanism that was not geared towards key exchange. Some attempts have been made since to define general requirements on an authentication mechanism which could serve as the basis for a formal proof. Cheng and Gligor have again formalised the goals of key exchange mechanisms [3] and, justifiedly, their attempt has been critizised for being too specific [13]. Other authors take the view that the goals of an authentication mechanism depend on its applications and that different claims will be made for different mechanisms [4, 13]. Before presenting our own suggestion, we will review two mechanisms proposed at some stage for inclusion in ISO/IEC 9798-3.

3 An example - the Canadian Attack

As so often, the mechanism subject to the Canadian Attack, to be discussed below, started its life as an improvement to a previous suggestion. The original is the following three pass authentication mechanism between two users A and B [6]. In the slightly simplified form given here, the values R_A and R_B are random challenges while S_A and S_B are the signature algorithms of A and B.

R1. $A \rightarrow B$: $R_A||B||\text{Text2}||S_A(R_A||B||\text{Text1})$
R2. $B \rightarrow A$: $R_B||A||R_A||\text{Text4}||S_B(R_B||A||R_A||\text{Text3})$
R3. $A \rightarrow B$: $R_B||B||S_A(R_B||B)$

This mechanism was critized because in the third step, A has to sign a string which is known to B in advance. Even worse, B can choose the value of R_B and might therefore induce A to sign a message which B could use in another context to impersonate A. To avoid this problem, the following change was suggested in [7], where the mechanism was also simplified by reducing the number of signatures which have to be computed.

R1. $A \rightarrow B$: $R_A||\text{Text1}$
R2. $B \rightarrow A$: $R_B||A||R_A||\text{Text3}||S_B(R_B||A||R_A||\text{Text2})$
R3'. $A \rightarrow B$: $R'_A||B||R_B||\text{Text5}||S_A(R'_A||B||R_B||\text{Text4})$

Now, B does not know in advance which string will be signed by A. We quote from [7]: *The random number R'_A is needed to prevent B from obtaining a signature on a prepared message ...*

This change survived in a few revisions of [7], until an attack was found by the Canadian member body of ISO [8]. In the presentation of this attack, which incidentally is quite similar to the attack against the CCITT X.509 mechanism presented in [1], we assume for the sake of simplicity that all text fields are empty.

$C \rightarrow B$: R_A $\qquad\qquad\qquad\qquad$ C pretends to be A
$B \rightarrow C$: $R_B||A||R_A||S_B(R_B||A||R_A)$
$C \rightarrow A$: R_B $\qquad\qquad\qquad\qquad$ C pretends to be B
$A \rightarrow C$: $R'_A||B||R_B||S_A(R'_A||B||R_B)$
$C \rightarrow B$: $R'_A||B||R_B||S_A(R'_A||B||R_B)$

Here, C succeeds in convincing B that it is A by initiating authentications with A and B under wrong names and using the string obtained as A's reply to its challenge as the third message in the authentication with B. The underlying flaw in this mechanism is again the missing link between the first and third message.

The subsequent modification, which now is part of the standard [7], remedies the situation as follows.

R1. $B \rightarrow A$: $R_B||\text{Text1}$
R2. $A \rightarrow B$: $R_A||R_B||B||\text{Text3}||S_A(R_A||R_B||B||\text{Text2})$
R3. $B \rightarrow A$: $R_B||R_A||A||\text{Text5}||S_B(R_B||R_A||A||\text{Text4})$

Now, B cannot control the entire message signed by A in the last step but the content of the message can be known to B in advance.

Could the underlying flaw in the previous version have been detected using the BAN logic? As stated above, the examples given in [1] are of little immediate help in approaching this question. To adopt the BAN logic for a formal analysis, we first have to specify the goals of authentication and then to examine whether the logic contains adequate operators to facilitate their statement. We may note at this stage that the BAN notation only captures encryption but neither signatures nor digital checksums. Public key encryption was added to BAN in [10]. Looking a step ahead, we may mention that Gaarder and Snekkenes [4] in their analysis of an authentication mechanism from CCITT X.509 introduce a special symbol for the notion of 'recipient' of a message. There, $\mathcal{R}(U_i, X)$ means that "U_i is the intended recipient of message X". Also, there are notations capturing the lifetime of messages and an attempt at distinguishing between freshness and recency of messages.

4   Proof obligations

The requirements on authentication mechanisms have been quoted in Section 1. These statements refer to a mechanism as a whole, not to its constituent messages. However, any belief the participants in an authentication exchange acquire will be based on the messages they have sent and received. Thus, we have to rephrase the general requirements in terms of the messages in an authentication mechanism. When we deal with challenge/response (handshake) mechanisms to prevent replay attacks, the following requirements seem to be appropriate for mutual authentication between two entities.

- Both entities believe that they have received a proper reply to their own challenge.

- Both entities believe that they have not replied to a replayed challenge.

The reader who, like the author, prefers to see authentication mechanisms being executed by nodes in a computer network rather than by humans, may replace 'believes' by 'accepts' in this context. The two requirements can be further refined. After going through the steps of an authentication mechanism, the entities A and B should hold the following beliefs.

(1) A believes: B has replied to my challenge.

(2a) A believes: I have replied to B's challenge.

(2b) A believes: B's challenge is fresh.

(3) B believes: A has replied to my challenge.

(4a) B believes: I have replied to A's challenge.

(4b) B believes: A's challenge is fresh.

When we define R_A and R_B to be nonces generated by A and B respectively and follow present usage in BAN, these conditions can be formulated as

(1) A believes B believes my nonce: $A \models B \models N_A$.

(3) B believes A believes my nonce: $B \models A \models N_B$.

However, this notation makes use of the $\models$-operator in a rather contrived fashion. If no new operator should be introduced, the concept of a "fresh user" is preferable. The interpretation of $\sharp B$ is then that B is the actual partner in a conversation, not a third party replaying messages originally generated by B. In that case we have

(1) A believes B is fresh: $A \models (\sharp B)$.

(3) B believes A is fresh: $B \models (\sharp A)$.

Alternatively, a "reply" operator could be introduced.

In stating condition (1) we have taken for granted that A believes B has replied to a fresh challenge as the challenge is generated by A. When we address our second requirement, it becomes useful to treat belief in the source of a challenge and belief in the freshness of a challenge as separate issues. As we will show, deficiencies such as those having led to the Canadian Attack can be exposed by an attempted proof of conditions (2) and (4). Conditions (2a) and (2b) could be captured by the "has jurisdiction" operator:

(2a) A believes B has jurisdiction over the challenge I replied to:
$A \models B \mapsto N_B$.

(4a) B believes A has jurisdiction over the challenge I replied to:
$B \models A \mapsto N_A$.

Conditions (2b) and (4b) can be expressed within the present framework of the BAN logic using the "freshness" operator.

5   Analysis

When we try to prove conditions (1)-(4) for the 'improved' mechanism, we find that after the last step B has no reason to believe that R_A originated from A. The first message R1 can be generated by anyone, the confirmation of the origin of R_A given in the original mechanism in R1 has been eliminated in the modified version. Therefore the basis for the proof of condition (4a) has disappeared. Conditions (1)-(3) can be proven. Note that we do not claim that (4a) cannot be proven and that we do not find the Canadian Attack. We only state that we do not see an obvious way for proving (4a) while this was quite easy in the original mechanism. In the original mechanism, condition (4b) could be shown by an (indirect) argument of the kind:

B believes: A has sent R3, therefore I have replied to a challenge R_A as expected by A, therefore R_A had been sent by A in a first message. Otherwise A would not be in step 2 of the mechanism expecting a reply to its fresh challenge R_A.

In the final standard, condition (2) holds because A receives a signed confirmation including its own challenge in message R3. Condition (4) holds because B receives a signed confirmation including its own challenge in message R2. In [11], some amendments to the second mechanism are suggested. The new third step

$$\text{R3''}. \quad A \to B : \ R'_A||\text{Text5}||S_A(R'_A||R_A||R_B||B||\text{Text4})$$

would again allow to prove condition (4) and add a field to the signed message which is not known to B in advance. In a scheme with message identifiers, [11], p.33,

$$\text{R1}. \quad A \to B: R_A$$
$$\text{R2}. \quad B \to A: R_B||\text{Text2}||S_B(\text{ID}_2||R_B||R_A||A||\text{Text1})$$
$$\text{R3}. \quad A \to B: \text{Text4}||S_B(\text{ID}_3||R_B||B||\text{Text3})$$

conditions (1)-(3) can be shown as above. Condition (4) follows again from an argument considering that A will send the third message only if it has reached step 2. If R_B is excluded from $R2$, a possibility considered in [11], then we cannot prove condition (2).

It may be seen as a disadvantage of message identifiers that principals have to keep track of the state they have reached in executing a mechanism. Introducing BAN rules for message identifiers requires a logic that can handle states. However, this may be required anyway, see e.g. [12] on the question of order of mechanism steps and termination of mechanisms. Also, one has to introduce rules for 'addresses' if mechanisms are used that abort when a message cannot be linked to the 'name' of the desired sender.

Acknowledgements

The author would like to thank Chris Mitchell for posing the question initiating this paper and for making available his most extensive collection of historic standards documents.

Bibliography

1. Burrows, M., Abadi, M. and Needham, R. (1990). A Logic of Authentication. DEC Systems Research Center, Report 39.

2. CCITT Draft Recommendation X.509 (1987). *The Directory Authentication Framework*, Version 7.

3. Cheng, P.-C. and Gligor, V.D. (1990). On the formal specification and verification of a multiparty session protocol. *Proceedings of the 1990 IEEE Symposium on Research in Security and Privacy*, pp.216–233.

4. Gaarder, K. and Snekkenes, E. (1990). On the formal analysis of PKCS authentication protocols. *Proc. Auscrypt'90*, J. Seberry and J. Pieprzik (eds), Springer LNCS 453, pp.106–121.

5. International Organization for Standardization, Genève, Switzerland (1991). ISO/IEC 9798-1 *Information technology – Security techniques – Entity authentication mechanisms; Part 1: General model.*

6. International Organization for Standardization, Genève, Switzerland (1990). ISO/IEC JTC1/SC27/WG2 N34 *Information technology – Security techniques – Entity authentication mechanisms; Part 3: Entity authentication mechanisms using a public key algorithm.*

7. International Organization for Standardization, Genève, Switzerland (1991). ISO/IEC JTC1/SC27/WG2 N51 *Information technology – Security techniques – Entity authentication mechanisms; Part 3: Entity authentication mechanisms using a public key algorithm.*

8. International Organization for Standardization, Genève, Switzerland (1991). ISO/IEC JTC1/SC27 N313 *Information technology – Security techniques – Summary of voting on Letter Ballot No.6, document SC27 N277, CD 9798-3.3 "Entity authentication mechanisms; Part 3: Entity authentication mechanisms using a public key algorithm".*

9. International Organization for Standardization, Genève, Switzerland (1993). ISO/IEC 9798-3 *Information technology – Security techniques – Entity authentication mechanisms; Part 3: Entity authentication mechanisms using a public key algorithm.*

10. Gong, L., Needham, R. and Yahalom, R. (1990). Reasoning about belief in cryptographic protocols. *Proceedings of the 1990 IEEE Symposium on Research in Security and Privacy*, pp. 234–248

11. Mitchell, C. and Thomas, A. (1993). Standardising authentication protocols based on public key techniques. *Journal of Computer Security*, *2*, 23–36

12. Snekkenes, E. (1991). Exploring the BAN approach to protocol analysis. *Proceedings of the 1991 IEEE Symposium on Research in Security and Privacy*, pp.171–181

13. Syverson, P. (1991). The use of logic in the analysis of cryptographic protocols. *Proceedings of the 1991 IEEE Symposium on Research in Security and Privacy*, pp. 156–70

A Galois theory of local reasoning in control systems with compositionality

M. Ingleby

Safety Critical Systems Unit, British Rail Research, Derby, and School of Computing and Mathematics, University of Huddersfield

Abstract The situated automaton concept is illustrated using two examples of open-loop control systems - a solid state interlocking (SSI), controlling signals and routes in an area of the railway; and a planning and scheduling system (PASS), providing decision support to a manufacturing operation. The essential geography of both types of system is reduced to an incidence relation and its associated Galois connection - between a set of controls and a set of features or attributes of the controlled environment. The Galois connection is related to similar connections used in formal concept analysis (FCA). In contrast to general FCA, the connections arising in the geographic data of the above automata are shown to possess a compositionality property K, which defines topology-like structures in the object and attribute sets of the automata. The multiple connectedness of these 'topologies' is exploited to partition a complex system into weakly-interacting parts or localities. These support a local reasoning process that can be used to deliver proof of safety, optimality, ...'correctness' of a system with great combinatorial efficiency.

1 Introduction: situated automata and their properties

Roughly speaking, a situated automaton is a state machine that monitors and controls an environment. It does so by maintaining in its state space a model of the environment to which it is linked by inputs and outputs. In general automata theory (for example, [12]), an automaton may be specified as a Mealy machine with input and output alphabets A and B and a state space S, as a state transformer with only an input alphabet and a state space, or in other ways. A specification that emphasises the accessibility relations between states, the so-called Büchi machine, is discussed elsewhere in these proceedings [4].

The different styles of specification are interchangeable. For example, if a Mealy machine, subject to an arbitrary input $a \in A$, undergoes changes of state and associated outputs described by next-state and output functions

$$\nu : A \times S \to S, \quad \text{with maplets} \quad (a,s) \mapsto v(a,s)$$

and

$$\omega : A \times S \to B, \quad \text{with maplets} \quad (a,s) \mapsto \omega(a,s)$$

then an equivalent specification of the same automaton as a state transformer is the function

$$\mu : A \times S \times B \to S \times B,$$

with maplets

$$(a,s,b) \mapsto \mu(a,s,b) \equiv (\nu(a,s),\omega(a,s)).$$

This remains true for situated automata. For the sake of simplicity in this article, we focus on Mealy machines.

With the above simplification, a *situated* automaton is a Mealy machine whose state space S models a controlled environment. The environment is generally represented as a function from a set of *elements* to a set of *values*. If, for example, the environment is a hydraulic plant controlled by flow-meters and valves, the elements are the meters and valves (or more generally the sensors and actuators) and the values are the possible sensor readings and actuator settings. The state of such an environment is a function from elements to values, and can be specified as a set of maplets assigning a value to each element. Any situated automaton controlling this environment has a similar set of maplets in its state space S - this being the automaton's *image* of its environment. The image is updated as the automaton receives inputs from alphabet A, whose letters are made up of element-to-value maplets. And the environment is changed by outputs from alphabet B whose letters are also made up of maplets of the same sort.

The class of situated automata considered in this paper consists of those whose element set is the union of two disjoint parts: a subset Φ of monitored environmental *attributes*, and a subset Γ of control *objects*. In the case of the hydraulic plant controller above, the attributes are the sensors and the objects are the actuators. More generally, attributes are elements which figure in the maplets used to build alphabet A, and objects are the elements figuring in the maplets defining alphabet B. We suppose that behavioural requirements to be placed on an automaton - safety, optimality, etc. - can be expressed using two forms of state predicate, P_1 and P_2:

- direct form, using a one-place predicate of state, $P_1(s)$ - meaning *'the current state is safe (or optimal, or correct in some other sense)'*;

- indirect form, using a two-place predicate of state and output, $Q(s, b)$, to define $P_2(s)$ as the sentence $\forall a \bullet Q(s, \omega(a, s))$ - meaning *'the current state outputs safely (or optimally, or correctly)'*.

The supposition is not very restrictive, and in the case of the hydraulic plant controller, for example, includes in type P_1 all restrictions on sensor readings and current actuator settings. Type P_2 includes all control principles that relate alterations in actuator settings to current sensor data.

The full expression of behavioural requirements of situated automata also makes use of the notion of Kleene closure ([20], or, for example, [10]). The closure of Mealy machine (A, B, S, ν, ω) is the machine $(A^*, B^*, S, \nu, \omega^*)$ with extended alphabets

$$A^* \equiv \emptyset \cup A \cup (A \times A) \cup (A \times A \times A) \cup \cdots, \quad B^* \equiv \emptyset \cup B \cup (B \times B) \cup (B \times B \times B) \cup \cdots$$

The strings of states in $S^* = S \cup (S \times S) \cup (S \times S \times S) \cup \cdots$ are used to define extended next-state and output functions $\nu^* : A^* \times S \to S^*$ and $\omega^* : A^* \times S \to B^*$ by recursion on the lengths of input strings:

$$\nu^*(\emptyset, s) \equiv s, \nu^*(a, s) \equiv (s, \nu(a, s))$$

and

$$\nu^*((a_1, a_2, \cdots a_n), s) \equiv (\nu^*((a_1, a_2, \cdots a_{n-1}), s), \nu(a_n, S_{n-1}));$$

$$\omega^*(\emptyset, s) \equiv \emptyset, \omega^*(a, s) \equiv \omega(a, s)$$

and

$$\omega^*((a_1, a_2, \cdots a_n), s) \equiv (\omega^*((a_1, a_2, \cdots a_{n-1}), s), \omega(a_n, s_{n-1}))$$

- where s_{n-1} is the last state of the string $\nu^*((a_1, a_2, \cdots a_{n-1}), s)$ reached from s under the action of input string $(a_1, a_2, \cdots a_{n-1})$. Although strings - such as input string $(a_1, a_2, \cdots a_n)$, output string $\omega^*((a_1, a_2, \cdots a_n), s)$ or state string $\nu^*((a_1, a_2, \cdots a_n), s)$ - are not sets, for convenience we shall allow the use of the set membership symbol $\in$ to denote the occurrence of terms in such strings. With this usage, the behavioural requirement "all states that are accessible from initial state s are correct and output correctly" takes the form

$$\forall x \bullet \forall s_k \bullet x \in A^* \wedge s_k \in \nu^*(x, s) \to P_1(s_k) \wedge P_2(s_k)$$

Elsewhere in these Proceedings [4], a direct attack on the proof of a requirement of this type is described - in a Büchi machine formulation.

Here we are concerned with an indirect attack, motivated by the fact that in many control systems, reinitialisation to different start states s is necessary for maintenance. Rather than proving by induction that all accessible states are correct, we seek to prove instead a slightly stronger requirement which guarantees not only the above but also a certain robustness of correct behaviour to reinitialisation. The stronger property was introduced in SAFECOMP proceedings [15] where it was called *safety-transitivity*. Here, we emphasise that other requirements than safety may also be involved, and speak simply of transitivity. An automaton is defined to be *transitive* if every state which is correct and outputs correctly has all its immediate successor states similarly correct :

$$\forall s \bullet \forall a \bullet (s \in S) \wedge P_1(s) \wedge P_2(s) \wedge (a \in A) \rightarrow P_1(\nu(a,s)) \wedge P_2(\nu(a,s)).$$

It is a matter of elementary induction on the lengths of input strings to prove

Proposition 1. *In a transitive automaton, every state accessible from a correct state is correct, that is*

$$\forall s \bullet \forall x \bullet (s \in S) \wedge P_1(s) \wedge P_2(s) \wedge (x \in A^*) \rightarrow P_1(\nu^*(x,s)) \wedge P_2(\nu^*(x,s)).$$

In the next section, examples are given of situated automata which model control and decision support systems of industrial interest. The examples indicate that the size of state space is a very significant barrier to machine assisted investigation of behavioural correctness requirements. The quantification over all states - an essential part of proof-goal sentences such as 'all accessible states are correct' or 'the automaton is transitive' - makes the problem of machine assisted proof of such requirements NP hard. If, for example, the controlled environment has M elements, each with only two values, the state space of any automaton which maintains an image of this environment in memory has 2^M states. Hence, any search of state space, in pursuit of correctness or transitivity goals, will take a time which is exponential in the size of the controlled environment.

In the rest of this paper, we propose a method of partitioning a large environment with M elements into a number of more tractable localities $L_1, L_2, \cdots, L_K$, with respective numbers of elements $m_1, m_2, \cdots, m_K$. The method takes into account the knowledge of which control objects act on which environmental attributes, capturing this knowledge as an incidence relation between the control elements in Γ and environmental attributes in Φ. A sufficient condition for the partitioning to work is that the environment has a compositionality property which we call K - because it equates to there being something akin to a Kuratowski closed set topology in Γ.

The mechanics of the partitioning method are founded on the theory of Galois connections, a branch of universal algebra which has come recently into widespread use for the purpose of exploratory data analysis ([7], [28], [29]).

The paper examines the clustering of states of an automaton with property K in a locality L_k, using an equivalence relation which defines two states s_i and s_j as locally equivalent in L_k if and only if the values of the two states coincide on the elements in L_k. A cluster dynamics is derived from the next-state function of the automaton, and the transitivity properties of automata are expressed as constraints on the cluster dynamics, to be proved by a combinatorially efficient form of 'local reasoning'.

2   Two examples of situated automata

To gain an idea of the breadth of applicability of the situated automaton concept, and of the combinatorial complexity of correctness proof, two very different examples are given below. The **first** models the closed loop control system which ensures the safe setting of railway signals in response to environmental data concerning the location of trains, the direction of points, the state of the signalman's control panel, etc. In this example, the correctness of strings of states lies in their conformance with historical safe signalling principles. The **second** example models the type of decision support that is used in some computer-aided manufacturing systems. These are open loop systems guiding human decision-makers in the scheduling of jobs and/or the planning of operations. For such decision-support systems, the correctness of states is defined via notional costs of operations along manufacturing paths, and a correct path is one having some local or global cost optimality.

Example 1. *A simple case of a signal controller can be described with reference to Figure 1, which is the scheme plan for a railway signal controller for an uncomplicated track layout. The physical attributes are track-sections (TA, TB, $\cdots$ TF - whose values may be occupied or clear - o or c) and points (P1, P2 - whose values may be controlled normal, controlled reverse, controlled or free to move normal, controlled or free to move reverse - cn, cr, cfn or cfr). The coarsest control objects are routes between signals (for example, R10A from signal S10 out through branch A of points P1 to signal S13 - whose values may be set or unset, s or xs). Then at a finer level there are subroutes U, of given direction over given tracks (for example, UA-CA over track TA in the direction from C to A) - whose values may be locked or free, ℓ or f. In a real signal controller, properly called a solid state interlocking (SSI) [5], slightly more refined control objects are added to those above, but this simple case captures the most important characteristics of SSI. The essentials of a scheme plan describing the environment*

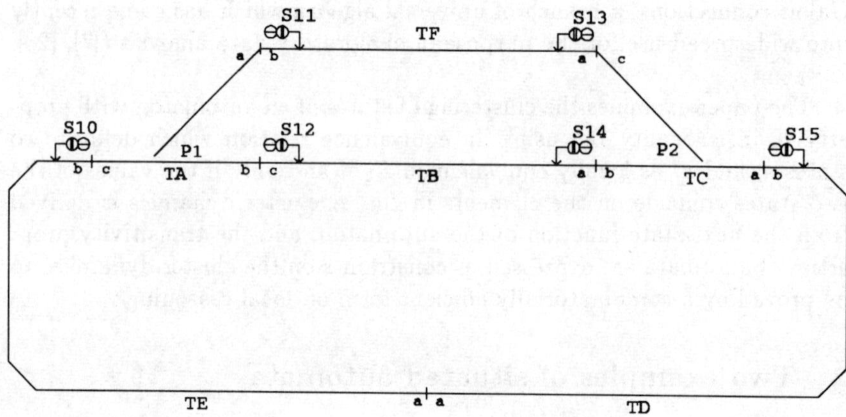

Figure 1. A signalling scheme plan

controlled are coded in a geographic database which is consulted in each cycle of operation of the automaton.

During one cycle, the automaton:-

• processes an input telegram from trackside equipment, updating values of certain physical attributes

• executes some interlocking rules (exemplified below) designed to keep the railway safe and available to traffic

• responds to route requests from a signalman's panel (see below) in a way that refuses requests for routes that conflict with any already set

• finally sends an output telegram communicating updated values of control elements to the trackside equipment.

Typical interlocking rules for the above scheme plan are

$$UA - CA\ f\ \text{if}\ R10A\ xs,\ TA\ c\backslash\quad\text{and}\quad UF - BA\ f\ \text{if}\ UA - CA\ f,\ TF\ c\backslash.$$

The first is a command to free the subroute from extremity c to extremity a over TA, but only if route R10A from signal S10 to S13 is not set, and if no train is detected on track TA. {This frees the subroute after a train has passed down the route to TF}. The second rule frees the subroute UF-BA from extremity b to extremity a over TF if the upstream subroute UA-CA has already been freed and there is no train on TF. The general intention implicit in these rules is to free subroutes behind a passing train so that other routes become available as soon as it is safe to set them. A typical route request, activated by receiving from the signalman's panel a telegram containing request code *Q10A for the aforementioned R10A, activates the panel request rule

$$*Q10A \quad \text{if} \quad P1 \, crf, \, UA - AC \, f, \, UF - AB \, f$$

$$\text{then} \quad R10A \, s, \, P1 \, cr, \, UA - CA \, l, \, UF - BA \, l\backslash$$

The request is granted by changing the SSI image of its railway environment to set R10A, to control points P1 in the reverse direction and to lock appropriate subroutes over TA and TF - but only if P1 is free to move reverse and all conflicting subroutes (over TA and TF) are free. Subsequently the changes will be sent to the trackside via output telegrams. The general intention behind requests is to grant them only when to do so involves neither simultaneous locking of conflicting subroutes over the same track nor controlling of points under or in front of an approaching train.

The type of reasoning about control systems with which BRR is concerned [6], [16] relates to the question 'Is there an unforeseen sequence of inputs which can drive an interlocking from a safe state to one in which the safety intentions are not realised?'. BRR approaches this question using a computer-assisted, NP hard search for unsafe states, seeking to prove *safety-transitivity* - that if an interlocking is in a safe state, then its next state must also be safe ([15] and Section 5). It turns out - as is usual with situated automata - that the search problem is NP hard, search time growing exponentially with size of the environment.

Example 2. *The second example of a situated automaton, a decision support system for planning and scheduling the manufacture of complex items - such as vehicle transmission shafts - from blocks of material. These systems are not closed-loop like a real-time chemical process controller or an SSI, but offer computer-based planning and scheduling support (PASS) for decisions under a human planner's control. The exact form of PASS decision support is very variable. Some PASS systems support only pure job-shop scheduling (JSS): each item is viewed as a job to be finished along with other jobs, scheduling to make optimal use of concurrency of operations of different jobs on different machines ([1]; [22]). Other PASS systems only offer computer-assisted process planning (CAPP): each separate job is viewed as a sequence of possible manufacturing processes with alternative process paths to the end of the job [13]. Quite some time ago in Japan, these separate aspects of planning and scheduling were integrated into more comprehensive modelling of the factory floor and more flexible decision-making*

[17]. The benefits of integration and flexibility are now well established (for example, [19], [21]), but before the benefits can be reaped, ever larger NP hard search problems must be solved.

Whatever form of decision support is offered, manufacturing proceeds by removing volumes $V_1, V_2, \cdots$ with different machines or flexible manufacturing cells $M_1, M_2, \cdots$ - milling, grinding, turning, etc., as appropriate. The manufacturing aim is to expose the surface of intended items. The surface of an item or items is represented as a simplicial complex $C = S_1 + S_2 + \cdots + S_N$, whose constituent simplices $\{S_i | 1 \leq i \leq N\}$ share a common boundary with a removable volume. The physical attributes for this example are the surface simplices. The surfaces take on values indicating whether they have been exposed or whether they are still covered. The removable volumes take on values removed or not removed. If the automaton is to assist with the planning task over periods when machines need maintenance, or if it controls a job shop where several different items may be under manufacture concurrently, there are also the machines to consider. These may take on values available or not available.

It is convenient to combine volumes and machines to define the control objects of PASS automata. The simplest control objects of the PASS automaton are defined in terms of volume removal operations (V_j, M_{jk}), the removal of V_j using machine M_{jk}. The order in which these operations are performed is subject to constraint, a string $[(V_1, M_{11}), (V_2, M_{21}), \cdots, (V_j, M_{jk})...]$ which satisfies the constraints being declared to be *legal*. Legality of a string is defined largely by the requirement that, to remove V_j with M_{jk}, a certain clearway for M_{jk} must be included in the previously removed volume $V_1 \cup V_2 \cup \cdots \cup V_{j-1}$. All the removal volumes which meet this clearway may be said to *cover* V_j. An illustration of covering constraints is given in Figure 2, where the clearway for a conical borer M7 removing V4 meets V1 and V2. In some cases, however, it may be that certain volumes require not only clearway but also especially precise placement relative to features already machined. In Figure 2, for example, if drill-hole V5 is to be placed accurately below the centre of the bored cone V4, one would have to agree that V4 covers V5 (dotted arrow in the constraint graph). We shall not include this additional constraint in further discussions of the problem in Figure 2, however.

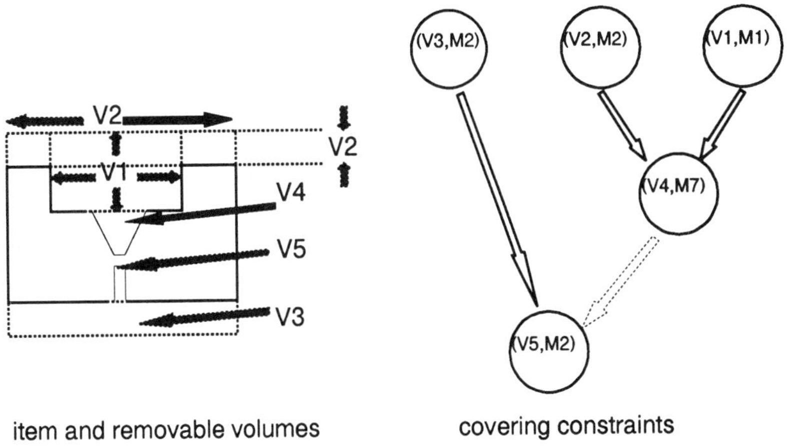

Figure 2. Sequencing of volume removals

The graph in Figure 2 is an example of the Hasse diagram of the pre-order relation *precedes*: the transitive closure of the covering relation. The rationale for precedence as transitive closure is that one must remove covers of covers, and their covers, and so on... to get clearway to remove a volume deep inside the source material of an item under manufacture. An operation (V_j, M_{jk}) is a control object of the PASS automaton analogous to the subroute of an SSI automaton. The analogue of a route is a compound operation conveniently conceptualised using the type of directed labelled graph illustrated in Figure 3. The nodes of the graph are functions of state of the automaton giving the set of volumes which are removed in a current state. The directed arcs are labelled with a removal operation (V_j, M_k). A mother node $\{V_1, V_2, \cdots, V_{j-1}\}$ is transformed along (V_j, M_{jk}) to daughter node $\{V_1, V_2, \cdots, V_{j-1}, V_j\}$ having the additional volume V_j removed. The arc is legal if the volume appearing in the arc label is such that all its predecessors on the precedence Hasse diagram are subsets of the union of volumes removed at the mother node. A compound arc or *path* is legal if each of its constituent arcs is legal, and these legal paths make up the set Γ for the PASS automaton. In Figure 3, the part of the graph of legal paths, those beginning with the removal of V1, is shown for the removal task defined in Figure 2.

During one cycle of the decision process, a PASS automaton:-

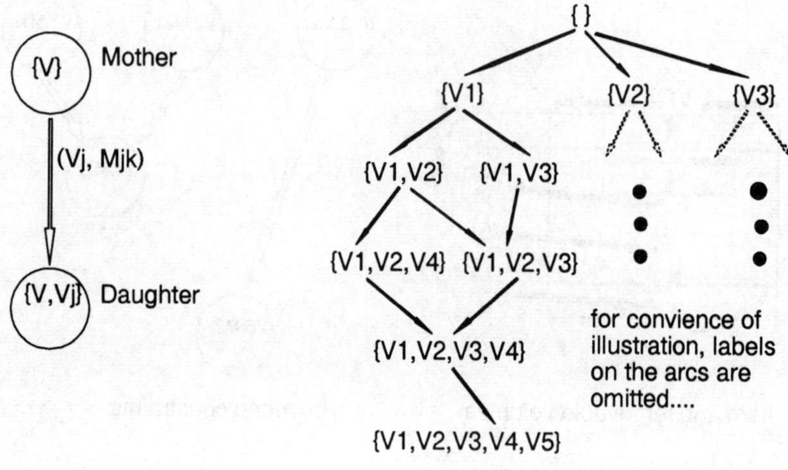

Figure 3. Graph-theoretic description of paths

- receives progress data updating the current state of surface exposure and volume removal of the items under manufacture, the availability of machines etc.

- applies decision procedures intended to identify an optimal way for the manufacturing system to follow legal paths for the jobs already set

- offers a menu of further jobs for the planner to set

- outputs timely decision support in the form of 'optimal' legal paths from the current state to completion of current jobs.

The quickest and simplest built-in decision procedures are nothing more than rules of thumb which are thought by experienced planners to tend towards low-cost manufacturing pathways through the legal path graph. For example [19]:-

- the least slack time rule {choose to work on the job for which the difference between expected remaining process time and the time to a completion deadline is least}

- the shortest imminence rule {having chosen a job, choose the operations which finish it soonest/cheapest on the available machines}.

A PASS automaton using these rules is inflexible in separating the choice of job (the JSS task) from the choice of processes to finish a job (the CAPP task). But besides putting aside the benefits of flexible Japanese-style task integration, such rules do not invariably drive a manufacturing system down an optimal legal path. Exhaustive search for an optimum can be better, but not unexpectedly, the exhaustive search problem is NP hard. Even without flexible integration, exhaustive search of the separate JSS and CAPP problems remain NP hard (see [3], and the references cited therein). A strictly correct PASS automaton will employ exhaustive search to output globally optimal paths from current states to completion of current jobs. If the state is revised as operations are completed and new jobs are set, and for any revision the search outputs a new path to completion which is optimal, then the automaton is transitive. A planner advised by a transitive automaton would be able to keep his manufacturing system always on an optimal path.

There is a considerable interest in devising parallel processing algorithms able to search a large legal path graph quickly enough for transitive decision-making. Other approaches not committed to parallel processing have emerged, too. Faced with decision problems too complex to solve in a timely way, most researchers have favoured faster decision procedures exploring some restricted neighbourhood of a promising path in the legal path graph. Genetic algorithms have been developed which treat paths as members of an ecological population able to spawn new paths like them, the least costly of the paths in the neighbourhood being favoured in the spawning of new paths. For a detailed review, see [24], and [14]. Simulated annealing has also been popular (for example, [25]). In the SA approach, a neighbourhood of progressively decreasing size is searched, imitating the shrinking state space available for ionic changes of configuration in an alloy as it anneals at progressively decreasing temperatures. The hope is that the broad early stages of search locate the part of state space where the smallest local cost minimum is to be found, while finer subsequent stages move in to this local minimum. In the physical world, the evidence of metastable states of matter and frozen in disorder shows that the hope is often too sanguine. Nevertheless, in operations research these approximate local search procedures often help planners to improve cost-effectiveness, but without achieving optimality or giving any guarantees of transitivity of performance in a changing task environment.

In this paper we are concerned to find ways of performing exhaustive search locality by locality - thereby achieving optimality and guaranteeing transitivity. We use a definition of locality which is derived from incidence relations, and founded on the theory of Galois connections. And we propose a rigorous local reasoning method from which global optimality, transitivity, and so on can be inferred, arguing that local reasoning is combinatorially less explosive than exhaustive global search.

3 Incidence relations, Galois connections and closed set topology

The SSI automaton's image of its railway environment is affected by the geography of routes over tracks in its controlled area. A route or subroute passes over certain tracks and points, and 'passing over' defines an incidence relation $R \subseteq \Gamma \times \Phi$. This can be generated automatically from a CAD-CAM map of the (railway) environment. The incidence relation defines a *Galois connection*, which is a pair of dual mappings $A \mapsto A^\perp, B \mapsto B^\perp$ - transforming respective subsets $A \subseteq \Gamma$ and $B \subseteq \Phi$ to $A^\perp \subseteq \Phi$ and $B^\perp \subseteq \Gamma$ defined by

$$A^\perp \equiv \{b \in \Phi | \forall a \in A \bullet \neg((a,b) \in R)\}, \qquad B^\perp \equiv \{a \in \Gamma | \forall b \in B \bullet \neg((a,b) \in R)\}.$$

A set B of tracks and points has an orthocomplement $B^\perp$ consisting of the routes and subroutes which do not pass over any elements of B; and the bicomplement $A^{\perp\perp}$ is the set of tracks and points not passed over by any of the routes and subroutes which do not pass over any elements of A. For the sake off simplicity in the rest of this paper, further discussion of this Galois connection will concentrate on *routes*. The task of adding constituent subroutes is a matter of simple routine. Table 1 gives the simplified incidence relation for the signalling area of Figure 1.

Similar incidence relations with their derived Galois connections are associated with PASS automata. The largest incidence relation of interest - that of legal paths on the surface simplices exposed by the operations along the path - is made unwieldy by the combinatorial complexity of legal path space. Therefore a smaller, more tractable relation obtained by limiting the control objects to simple arcs (V_i, M_{ik}), is studied. For the smaller relation, Γ is the set of operations or simple arcs, while Φ is the set of surface simplices. To take into account the precedence relation, denoted $\geq$, between removal operations, the smaller incidence relation is compiled recursively using

Definition 2. *Operation (V_i, M_{ik}) is incident on surface simplex S_j iff either $S_j \subseteq \partial V_i$ {direct incidence} or*

$$\exists (V_{i'}, M_{i'k}) \bullet (V_i \geq V_{i'}) \wedge (V_{i'} \text{ is incident on } S_j)$$

{indirect incidence}. {Less formally, the operation covers the simplex if it has to be carried out to expose the simplex}.

Table 1. Incidence of routes on points and tracks

ATTR→ OBJ↓	P1N	P1R	TA	TF	TB	TE	TD	TC	P2N	P2R
R10A	0	1	1	1	0	0	0	0	0	0
R10B	1	0	1	1	0	1	0	0	0	0
R11	0	1	1	0	0	1	1	0	0	0
R12	1	0	1	0	0	1	1	0	0	0
R13	0	0	0	0	0	1	1	1	0	1
R14	0	0	0	0	0	1	1	1	1	0
R15A	0	0	0	0	1	0	0	1	1	0
R15B	0	0	0	1	0	0	0	1	0	1

Table 2. Incidence of operations on surface simplices

ATTR→ OBJ↓	S_1	S_2	S_3	S_4	S_5
(V_1, M_1)	1	0	0	0	0
(V_2, M_2)	0	1	0	0	0
(V_3, M_2)	0	0	1	0	0
(V_4, M_7)	1	1	0	1	0
(V_5, M_2)	0	0	1	0	1

The results of compiling this shorter relation for the volume removal problem of Figure 2 are given in Table 2. The compilation was carried out using the clearway constraints alone - ignoring the dotted arrow in Figure 2. {It could have been carried out *with* the dotted arrow, but this would not illustrate the locality concept nearly so well!}. The labelling of simplices used in the table is such that S_i is the segment of the surface under manufacture which forms part of the boundary ∂V_i. The Galois connection $A \mapsto A^{\perp}, B \mapsto B^{\perp}$ associated with the compiled relation R has the following interpretation. If $B = S_1 + S_2 + \cdots + S_N$ is a segment of the surface of an item or items under manufacture, the orthocomplement $B^{\perp}$ is the set of operations which do not cover B. And the bicomplement $B^{\perp\perp}$ is the enlarged segment of the surface under manufacture not covered by operations that do not cover B.

Galois connections like these are quite pervasive in traditional mathematics, and broadly speaking come in two flavours - algebraic and topological. In the linear *algebra* of finite- dimensional vector spaces, for example, if Γ is the set of points in space $\mathcal{E}$, and Φ is the dual space $\mathcal{E}^*$ of functionals on ε. Then functional $\phi \in \mathcal{E}^*$ is said to be incident on a vector $a \in \mathcal{E}$ iff $\phi(a) \neq 0$. The orthocomplement $A^\perp$ of $A \subseteq \Gamma$ is the so-called annihilator of A, the subspace of the dual space $\mathcal{E}^*$ consisting of functionals which vanish on A; and bicomplement $A^{\perp\perp}$ is the so-called linear closure of A, the smallest subspace of $\mathcal{E}$ which includes A. By way of contrast, in general *topology* [18], if Γ is the set of points in a topological space and Φ the topology of open sets of that space, an open set V of the topology Φ is said to be incident on point $a \in \Gamma$ iff $a \in V$. Then orthocomplement $A^\perp$ of subset $A \subseteq \Gamma$ is the collection of all open sets disjoint from A, and bicomplement $A^{\perp\perp}$ is just the topological closure of A. Galois theorists refer to bicomplements as closures, and call a set which is equal to its own bicomplement a *closed set*. The set complement of a closed set may be called an *open set*, and this definition of *open* coincides with the topological definition in the case of the connection derived from the incidence of open sets on points of a topological space. It is not the case, however, that the open sets of objects defined by an arbitrary Galois connection *always* constitute a topology in Γ: in the algebraic example above, the closure of the empty set of points consists of the singleton point at the origin $\{0\}$ - and this closure would be empty in a true topology!

Despite such differences between Galois connections, a number of properties of orthocomplements are shared by all the above examples, topological and otherwise, and are indeed true of any incidence relation [7]:

Proposition 3. *For any incidence relation between non-empty sets Γ and Φ, empty subsets $\emptyset_\Gamma$ and $\emptyset_\Phi$, and arbitrary subsets A, A_1, A_2 in Γ and B, B_1, B_2 in Φ, have associated Galois orthocomplements which satisfy:*

$$\emptyset_\Gamma^\perp = \Phi, \qquad \emptyset_\Phi^\perp = \Gamma; \tag{3.1}$$

$$A_1 \subseteq A_2 \rightarrow A_2^\perp \subseteq A_1^\perp, \qquad B_1 \subseteq B_2 \rightarrow B_2^\perp \subseteq B_1^\perp \tag{3.2}$$

$$A \subseteq A^{\perp\perp}, \qquad B \subseteq B^{\perp\perp}; \tag{3.3}$$

$$A_1 \subseteq A_2 \rightarrow A_1^{\perp\perp} \subseteq A_2^{\perp\perp}, \qquad B_1 \subseteq B_2 \rightarrow B_1^{\perp\perp} \subseteq B_2^{\perp\perp}; \tag{3.4}$$

$$A^\perp = A^{\perp\perp\perp}, \qquad B^\perp = B^{\perp\perp\perp}; \tag{3.5}$$

$$\cap_{i\in I}\,(A_i^{\perp}) = \cup_{i\in I}(A_i)^{\perp} \qquad \cap_{i\in I}\,(B_i^{\perp}) = (\cup_{i\in I}B_i)^{\perp}. \qquad (3.6)$$

The dual of (3.6), obtained by changing the unions in (3.6) to intersections and vice versa, is not always true. All that can be said in general is

$$\cup_{i\in I}\,(A_i^{\perp}) \subseteq (\cap_{i\in I}A_i)^{\perp} \qquad \cup_{i\in I}\,(B_i^{\perp}) \subseteq (\cap_{i\in I}B_i)^{\perp} \qquad (3.7)$$

A case where the inclusion is strict is shown in Figure 4. Sets $X1 = \{x_1\}$ and $X2 = \{x_2\}$ are singleton sets of basis vectors from two-dimensional vector space $\mathcal{E}$. Then $X1^{\perp}$ and $X2^{\perp}$ are one-dimensional subspaces of $\mathcal{E}^*$, but $(X1 \cap X2)^{\perp} = \varnothing_{\mathcal{E}}^{\perp} = \mathcal{E}^*$, from which it follows that $X1^{\perp} \cup X2^{\perp} \subset (X1 \cap X2)^{\perp}$. Orthocomplementing the situation in Figure 4, $X1^{\perp\perp}$ and $X2^{\perp\perp}$ are the respective one-dimensional subspaces of $\mathcal{E}$ containing vectors x_1 and x_2, and $(X1 \cup X2)^{\perp\perp} = \mathcal{E}$ because it is the linear closure of a basis set. This yields another strict containment, $X1^{\perp\perp} \cup X2^{\perp\perp} \subset (X1 \cup X2)^{\perp\perp}$. The most that can be said in general about unions of closures is

$$\cup_{i\in I}\,(A_i^{\perp\perp}) \subseteq (\cup_{i\in I}A_i)^{\perp\perp} \qquad \cup_{i\in I}\,(B_i^{\perp\perp}) \subseteq (\cup_{i\in I}B_i)^{\perp\perp}, \qquad (3.8)$$

which the active reader will be able to infer by bicomplementing the set-theoretic truism

$$\forall i' \bullet A_i' \subseteq \cup_{i\in I}A_i.$$

Similar arguments with intersections of closures can be used to establish

$$\cap_{i\in I}\,(A_i^{\perp\perp}) \subseteq (\cap_{i\in I}A_i)^{\perp\perp} \qquad \cap_{i\in I}\,(B_i^{\perp\perp}) \subseteq \cap_{i\in I}B_i)^{\perp\perp}, \qquad (3.9)$$

with strict inclusion being a possibility in the case of Galois connections with an algebraic flavour.

The properties of the bicomplement or closure operators which distinguishes a Galois connection with an algebraic flavour from one with a topological are the Kuratowski properties [18]

$$\varnothing_{\Gamma}^{\perp\perp} = \varnothing_{\Gamma}, \qquad \varnothing_{\Phi}^{\perp\perp} = \varnothing_{\Phi}, \qquad (X_1 \cup X_2 \cup \cdots)^{\perp\perp} = X_1^{\perp\perp} \cup X_2^{\perp\perp} \cup \cdots$$

They are true of the topological example - in any topological space, the empty set and the whole space are closed, and the closure of any finite union is the union of closures of summands. The properties themselves imply that the closed sets of Γ and Φ form a closed set topology - a simple corollary of Proposition 3 (3.6) is that arbitrary intersections of closed sets are closed, and an equally simple consequence of the third Kuratowski property is that finite unions of closed sets are closed.

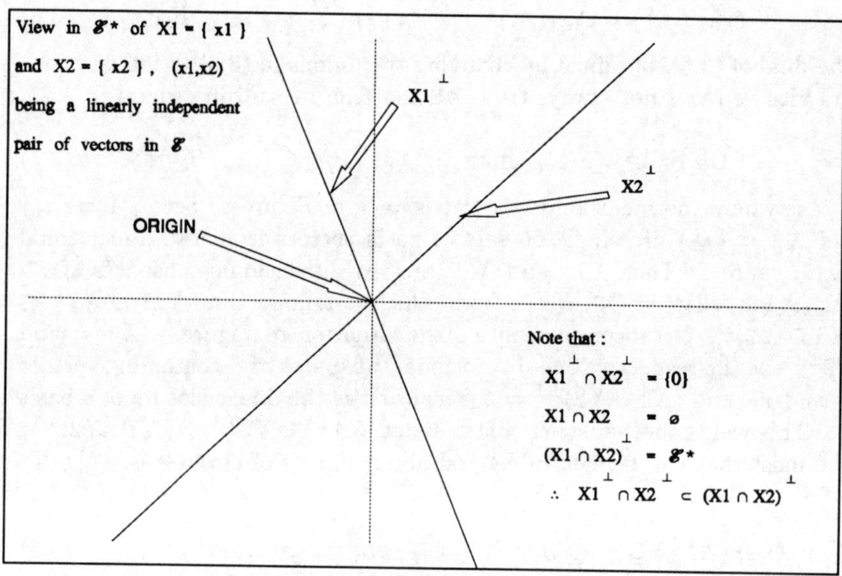

Figure 4. A union of closures strictly contained in linear closure of union

It was argued above that the closure of an empty set of vectors is not empty, in violation of the first and second Kuratowski properties. And the orthocomplementation the situation on Figure 4, used above to establish (3.8), provides an example of a union of closed, one dimensional subspaces contained in but not equal to closure of their union, in violation of the third Kuratowski property. In a SAFECOMP paper [15], it was argued that situated automata have incidence relations with the property K:

$$\forall g \in \Gamma \bullet \exists \phi \in \Phi \bullet (g \cdot \phi) \in R, \text{ and dually } \forall \phi \in \Phi \bullet \exists g) \in \Gamma \bullet (g, \phi) \in R.$$

Property K, which may be called compositionality, requires that every control object is incident on at least one monitored attribute, and every attribute has at least one control object incident upon it. In the paper we showed:

Proposition 4. *For any incidence relation with the property K, the closure operators $x \mapsto X^{\perp\perp}$ on both Γ and Φ satisfy the Kuratowski properties.*

The proof is not quite straightforward, but can be made so by first noting that (3.8) above implies that for any sets X_1 and X_2 - of objects or of attributes -the bicomplement mappings satisfy the set inequalities

$$X_1^{\perp\perp} \cup X_2^{\perp\perp} \subseteq (X_1 \cup X_2)^{\perp\perp}.$$

And then proving the following easy lemmas:

Lemma 5. *For any incidence relation, the bicomplements of empty sets of objects and of attributes are empty iff the relation has property K.*

Proof. Simple consequence of definition of orthocomplement and property K.

Lemma 6. *For any incidence relation with property K, and any sets X_1 and X_2 - of objects or of attributes - the bicomplement mappings satisfy the set inequalities*

$$(X_1 \cup X_2)^{\perp\perp} \subseteq X_1^{\perp\perp} \cup X_2^{\perp\perp}.$$

Proof. In the case where the X's are sets of objects,

$$(X)^{\perp\perp} = \{x | \forall y \bullet [\forall x_0 \bullet (x_0 \in X) \wedge \neg(x_0, y) \in R \to \neg(x, y) \in R]\}$$

has a contrapositive form

$$(X)^{\perp\perp} = \{x | \forall y \bullet [(x, y) \in R \to \exists x_0 \bullet (x_0 \in X) \wedge (x_0, y) \in R\}.$$

Then, for an arbitrary x from $(X_1 \cup X_2)^{\perp\perp}$, by property K there is a y is incident on x; and from the contrapositive form of the bicomplement there is an x_0 in (X_1, X_2) on which y is also incident; when $x_0 \in X_1, x \in X_1^{\perp\perp} \subseteq X_1^{\perp\perp} \cup X_2^{\perp\perp}$. Similarly when $x_0 \in X_2$, $x \in X_1^{\perp\perp} \cup X_2^{\perp\perp}$.

As a consequence of Proposition 4, the closed sets of objects and attributes constitute respective topologies in Γ and Φ. It turns out that the topology of objects can be used to partition the object sets of situated automata with property K into disjoint components or localities, as elaborated in Section 4.

4 Localities as both concepts and topological components

The general properties of Galois connections provide the basis for a modern theory of data analysis known as *formal concept analysis* (FCA), now widely used in the exploration of medical databases and psychometric data (for example, in [23]). It is argued below that the addition of the topological compositionality property K provides concepts sufficient to decompose NP-hard search problems in the state spaces of automata into more tractable parts.

In FCA a *context* consists of a pair (Γ, Φ) of sets and an incidence relation $I \subseteq \Gamma \times \Phi$. As in the examples above, the members of Γ are known as the *objects* of the context, and the members of Φ are known as its *attributes*. The localities which are needed to support local reasoning in situated automata are instances of what Rudolf Wille ([27]) called *concepts*. In general [11], a concept is a pair of sets *(A,B)* with $A \times B \subseteq \Gamma \times \Phi$ in which

1. the so-called *extent*, A, is the set of objects making up the concept and is closed for the Galois connection generated by $\neg I$, the set complement in $\Gamma \times \Phi$ of relation I

2. the so-called *intent*, B, is the set of attributes making up the concept, and is the orthocomplement with respect to $\neg I$ of A {and automatically closed by Proposition 3. (3.5)}.

The closure requirement, and the use of $\neg I$ is inspired by the observation that closure $A^{\perp\perp}$ is the set of objects which share all the attributes shared by the objects of A. Thus, an A which is closed is maximal with respect to the sharing of attributes. Wille and his collaborators have in fact combined the orthocomplement and set complement to associate with relation I a Galois connection $A \mapsto A', B \mapsto B'$ which is that generated by the orthocomplement mappings of $\neg I$.

The concepts in a given context can be ordered into a hierarchy by the object-inclusion relation $\leq$:

$$(A_1, B_1) \leq (A_2, B_2) \leftrightarrow A_1 \subseteq A_2 \quad \text{(or equivalently, } \leftrightarrow B_2 \subseteq B_1\text{)}.$$

There is a fundamental theorem of FCA ([7], [29]), which includes as a special case Dedekind's construction of the real number lattice from the order relation of the rational numbers, and is essentially an elaborate corollary of Proposition 3:

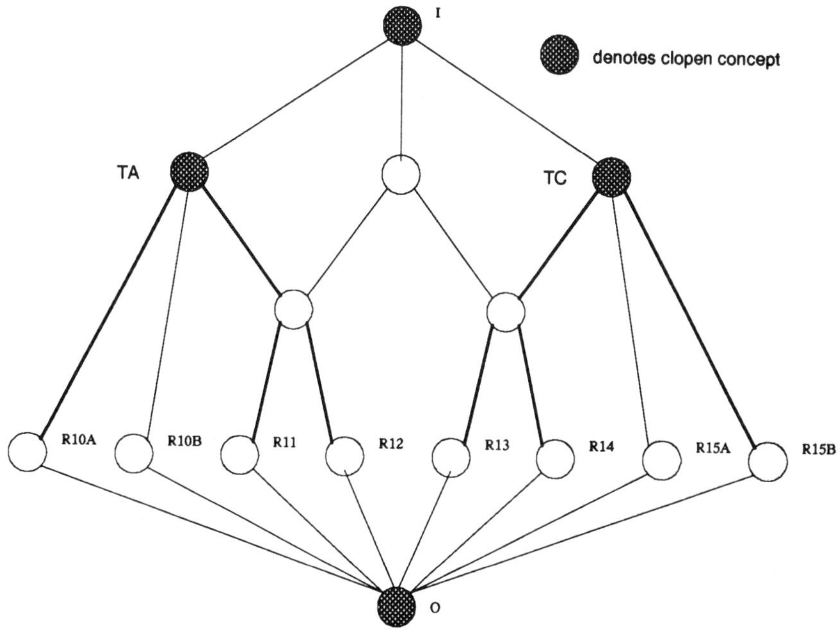

Figure 5. The concept lattice for Table 1

Concept Lattice Theorem. For any context $I \subseteq \Gamma \times \Phi$ the concepts with the above partial order form a complete lattice in which joins ('$\vee$') and meets ('$\wedge$') are given by

$$\vee_{j \in J}(A_j, B_j) \;\; = \;\; ((\cup_{j \in J} A_j)'', (\cap_{j \in J} B_j))$$

$$\wedge_{j \in J}(A_j, B_j) \;\; = \;\; ((\cap_{j \in J} A_j), (\cup_{j \in J} B_j)'').$$

The concept lattice of context $I \subseteq \Gamma \times \Phi$ is denoted $B(\Gamma, \Phi, I)$, and necessary and sufficient conditions for a complete lattice can be represented as a $B(\Gamma, \Phi, I)$ have been formulated ([7]). Much effort has been devoted to the automatic generation of concept lattices, and to their display as Hasse diagrams, using computers ([26],[28]). The most elementary way of generating the concept lattice of a relation is to form the closures of all subsets of Γ, starting with singleton subsets and then pairs, triples and so on. In this way one can find all extents and calculate the intent of each of the distinct extents. In practice, there is much duplication because different subsets can have the same closure. The elementary computation with all its duplication is of complexity $2^{|\Gamma|}$, and is a very inefficient form

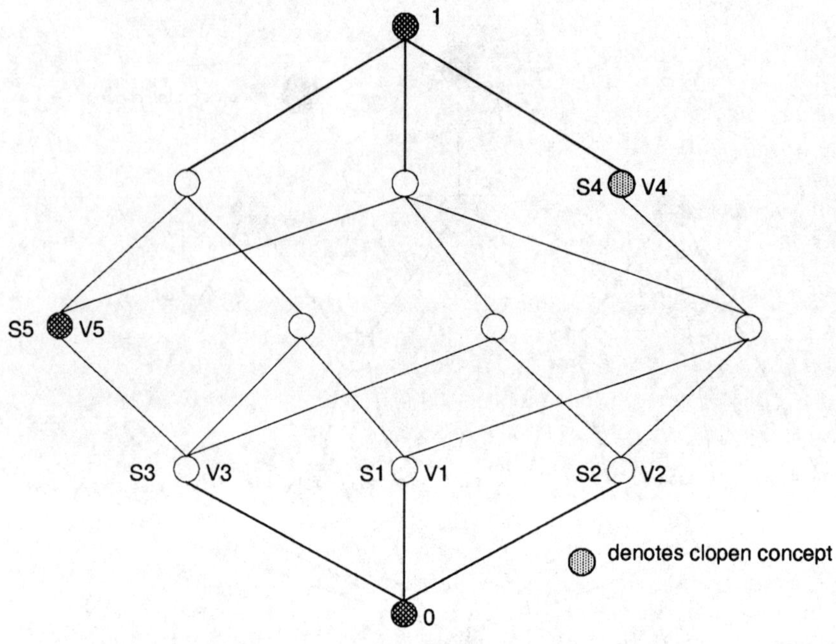

Figure 6. The concept lattice for Table 2

of knowledge compilation. The best closure-generating algorithm to date
is $0(|\Gamma|^2)$, developed by a member of Wille's group ([8],[9]).

The results of applying Ganter's algorithm to the contexts shown in Tables 1 and 2 are shown in Figures 5 and 6. In the figures, labels on the left of
concept nodes are intents, and labels on the right of nodes are extents. The
labelling convention used by the FCA community is that only nodes whose
extents or intents are closures of singleton sets are labelled. In the figure,
the concept label TA refers to the concept with intent $\{TA\}'' = \{TA\}$, inferred by noting that there are no attribute labels above TA in the Hasse diagram. The extent of node TA is the set of routes {R10A,R10B,R11,R12},
which can be inferred by noting the object labels on nodes below TA on
the Hasse diagram. The concept labelled TC, similarly, has intent {TC}
and extent {R13,R14,R15A,R15B}. The two extents are complementary,
$\{R10A,R10B,R11,R12\}^c = \{R13,R14,R15A,R15B\}$. Hence both extents
are clopen sets in the topology of routes.They constitute a partition of the
set of routes - into the cluster of routes which pass over TA and the cluster
passing over TB.

The lattice of Figure 6 also has a pair of concepts with complementary extents. The node labelled V5 on the right represents the concept with extent $\{V_5\}'' = \{V_3, V_5\}$, the right hand side if this equality being inferred by looking at labels on the nodes below 0 in the Hasse diagram. Wille's labelling convention guarantees that one can always infer the extent of a concept by looking in this way for the objects which label nodes below a given concept in the Hasse diagram. In Figure 6, label V4 denotes the concept with extent $\{V_4\}'' = \{V_1, V_2, V_4\} = \{V_5, V_3\}^c$ and intent $\{S_1, S_2, S_4\}$. The two intents partition the surface of the item into two localities or 'sides' whose production can be planned independently of each other.

It is generally possible to search a concept lattice for clopen sets as it is being generated by the Ganter algorithm, simply by adding, as each closed set E is formed, a check to see whether E^c, its set complement in Γ, is also closed. The clopen sets so generated arise in complementary pairs, the set complement of a clopen set being also clopen. It turns out that the clopen sets constitute a sublattice of $\mathbf{B}(\Gamma, \Phi, I)$ that may be denoted $\mathbf{C}(\Gamma, \Phi, I)$. The shaded circles of Figure 5 define $\mathbf{C}(\Gamma, \Phi, I)$ for the context of Table 1.

The atoms or irreducible elements of a $\mathbf{C}(\Gamma, \Phi, I)$ are the clopen concepts having no non-trivial clopen subconcepts, and it is generally possible to arrange some of these so that their extents form a partition of Γ. The procedure operates with a descending chain $\Gamma_0 \supset \Gamma_1 \supset \Gamma_2 \supset \cdots \Gamma_n \supset \Gamma_{n+1} \supset \cdots$ generated as follows:

1. begin with $\Gamma_0 = \Gamma$;

2. stop if Γ_n is irreducible;

3. otherwise choose any irreducible $E_n \subset \Gamma_n$, and let $\Gamma_{n+1} = \Gamma_n \cap E_n^c$ for a repeat of 2 with Γ_{n+1} in place of Γ_n.

The result is a family $(B_j)_{j \in J}$ of distinct concepts with $B_j = (E_j, E_j')$ whose extents form a partition of Γ. Such a family may be referred to as a *concept partition*.

Concept partitions are easily extracted from the concept lattice of an automaton, and the extents associated with any such family capture the notion of locality needed for local reasoning. In what follows, it is assumed that such a concept partition has been obtained, and the extents will be denoted L_j instead of E_j above - and referred to as localities.

5   Local clustering of states

During the search (for unsafe states of an SSI automaton, or for a minimal-cost path between a specified pair of nodes of a PASS automaton, or generally for incorrect states of any automaton), there is a need both to set up a search- or state-space and to define the objective of the search (using, for example, predicates of state).

In the case of SSI, states are functions, $s : \Gamma \times \Phi \to V$, mapping elements of the railway to their values. The state predicates P_i of Section 1 are sentences whose terms are the maplets which make up a state. An example of a state predicate which is important in safety reasoning is

$P_1(s)$, in state s, for all routes R, if R is incident on any track section T and its subroute U also incident on T is locked, then every other subroute U' incident on T is free.

In the case of PASS, states are paths between a given pair of nodes in the legal paths graph exemplified in Figure 3. State predicates are sentences referring to the operations on these paths. An example of a state predicate which is important when reasoning about the correctness of decision procedures is

$P_2(s)$, for all operations on path s, and for all legal paths s' with the same extremities as s, the total execution times $T(s)$ and $T(s')$ of operations on s and s' respectively, satisfy $T(s) \leq T(s')$. .

{Obviously execution times could be replaced by other cost-related objective functions reflecting the relative setting-up cost and running costs of operations on different machines}.

Predicates $P_i(i = 1, 2)$ can be localised to a fixed locality of clopen concept $B = (L, L')$ of the automaton:

$P_1^L(s)$, for all routes R *in locality L* of B, if R is incident on track-section T and if its subroute U over T is locked, then every other subroute over T is free;

$P_2^L(s)$, for all operations in locality L of B on path s, and for all legal paths s' with the same extremities as s, the total execution times $T^L(s)$ and $T^L(s')$ of operations in L and on s and s' respectively, satisfy $T^L(s) \leq T^L(s')$.

The general predicate form which supports this type of localisation is a conjunction of clauses in which an object is the leading term, and this form may be called localisable. For a predicate in *localisable* form, the conjunction over objects may be carried out locally for all localities in a partition. The form of P_1 obviously conforms to this pattern, and

$$\forall s \bullet (\forall j \bullet P_1^{Lj}(s)) \leftrightarrow P_1(s)$$

In the case of P_2, the additive nature of execution times allows one to infer global optimality from the conjunction of local optimalities:

$$(\forall j \bullet T^{Lj}(s') \geq T^{Lj}(s)) \rightarrow \sum_j T^{Lj}(s) \rightarrow T(s') \geq T(s).$$

The above arguments lead to

Proposition 7. *If $(B_j)_{j \in J}$ - with $B_j = (L_j, L'_j)$ - is a partition of the concept lattice* $\mathbf{B}(\Gamma, \Phi, I)$*, and P is a predicate of state in localisable form with localisations P^{Lj}, then*

$$\forall s \bullet (\forall j \bullet P^{Lj}(s)) \rightarrow P(s).$$

To appreciate the advantages of local proof, it must be noted that the Boolean value of a $P_i^L(s)$ depends only on the state values of objects in the locality L. This means that localised predicates depend on the class or cluster of states $[s]$ to which s belongs. The equivalence relation for locality L is:

(case of SSI) $s_1 \sim_L s_2 \leftrightarrow \forall e \in L \bullet s_1(e) = s_2(e)$
                     {routes take equal state values inside the locality};

(case of PASS) $s_1 \sim_L s_2 \leftrightarrow \forall(V, M) \in L \bullet V$ removed by machine M
                      on path $s_1 \leftrightarrow V$ removed by machine M on
                      path s_2
                      {the L-volumes on s_1 are all on s_2 where they are
                      removed by the same machines, and conversely}.

When the need arises to prove that properties of a situated automaton are invariant over all states, there is enormous combinatorial gain from proceeding locally, and dealing not with all states, but with all clusters of states *modulo* equivalence relations like the above. For example, when working with an SSI larger than that of Figure 1, to prove exhaustively that the geographic data explored by Atkinson and Cunningham [2] define a safety-transitive automaton, four localities were used. A PROLOG proof assistant performed the whole area proof in 75 minutes, but with the area partitioned into the four localities, the local proof with clusters of states took about a minute per locality with the same computer.

6   State transitions and non-deterministic cluster dynamics

To avoid giving a false impression, it should be emphasised that the combinatorial advantages of locality-based proof are bought at a price: clusters of states do not evolve deterministically. That is, for two possible states in a given cluster, the respective next states may be located in different clusters.

The following SSI example illustrates how dependencies between localities can give rise to non-deterministic cluster dynamics. The illustration makes use once more of the naming convention for subroutes in GDL: subroutes over track-section **TD** are named UD- with a pair of letters after the hyphen denoting the direction of the subroute. For example, UD-AB is the subroute from extremity a of TD to extremity b of TD (see Figure 1). Focussing on the locality labelled TA in Figure 5, with extent {R10A,R10B,R11,R12} and intent {TA}, suppose the system is in a state with locked subroutes UE-AB and UD-BA (for route R13 from S13 to S10, outside the locality, which is not set). Suppose further that R11 is not set and that during the current cycle of the interlocking, the following interlocking rules are to be processed:

UD-BA f if UB-BA f, UC-CA f, TD c
           {subroute release, freeing UD-BA iff TD is clear and
           its upstream subroutes over TC on R13 and R14 have
           been freed};

UE-AB f if UD-BA f, TE c
           {subroute release, freeing UE-AB iff TE is clear, and
           upstream subroute UD-BA above has freed first};

*Q11 if P1 crf, UA-CA f, UE-AB f, UD-BA f
      then R11 s, P1 cr, UA-AC l, UE-BA l, UD-AB l\
           {route request, setting R11, controlling P1 to the reverse
           direction, and locking subroutes on R11 iff the two
           opposing subroutes above have been freed and points P1
           are reverse or free to move reverse, as below};

*P1R if UA-CB f, UA-BC f, TA c\
           {points free to move test, succeeds iff TA is clear of trains
           and subroutes over TA requiring P1 to be in the normal
           direction are free}.

Clearly there are four possible cases for the locking of subroutes in the next state, depending on the occupancy of TD and TE - track-sections on which routes of two localities are incident. In three of the four cases, when at least one of these two track-sections is occupied R11 remains unset and its subroutes remain free. In the fourth case where TE and TD are both clear, the test in panel request *Q11 succeeds, and the command to set R11 and lock its subroutes is executed as the interlocking processes panel request *Q11 above. The next state in this fourth case lies in a different $\sim_{TA}$ state cluster from the next state in the other three cases, so two possible next clusters have to be considered in the cluster dynamics. This non-determinism arises because the behaviour of locality TA elements during the cycle depends on what is currently true of the TD concept which is outside the TA locality.

Obviously the same kind of non-determinism will arise in the PASS example, where the availability of machines to execute operations in one locality (set of jobs), depends on their use on operations removing volumes of other localities. It is not, however, a barrier to the automation of correctness proofs: all possible next clusters are readily enumerated, and the non-determinism increases by a marginal factor (3 or 4, in practice) the amount of correctness-checking of the local cluster dynamics.

For the automaton controlling Figure 1, with 6 track-sections, 2 points, 8 routes and 16 subroutes, there are 32 elements altogether, hence S has 2^{32} member states. Each must be checked for transitivity, execution time being proportional to 2^{32}. If, however, one reasons locally, partitioning the elements into localities TA and TC each with 4 routes, 1 set of points and 1 track-section over which all the local routes pass, then the local reasoning involves only 2^6 state clusters per locality. Of course the non-determinism of next-clusters makes local next- state correctness checking more complex than the global checking - but the complexity multiplication factor cannot exceed the number of ways of assigning values to track-sections and incident subroutes shared between the two localities (2^{12}).

More generally, if there are M elements and 2^M states, partitioned for local reasoning into localities $L_1, L_2, \cdots, L_J$ with respective numbers of local clusters $2^{m1}, 2^{m2}, \cdots, 2^{mJ} (mj's$ such that $\sum_{j=1}^{j=J} mj \leq M)$, then:-

1. without localisation the transitivity checking takes time proportional to 2^M

2. with localisation the execution time is proportional to $(2^{m1} + 2^{m2} + \cdots + 2^{mJ}) <<< 2^M$, though the proportionality constant is greater in this case.

This strong inequality shows the advantage of local reasoning - if one is attempting to engineer a computer assisted proof of correctness with localisable state predicates!

The complexity multiplication factor associated with the non-determinism of local cluster dynamics works against the inequality, and is difficult to estimate generally. For the examples explored to date, the factor depends both on the extent to which the original state dynamics is influenced by attributes of state on which objects from more than one locality are incident. If this extent is small, one might say that localities interact *weakly* - but a precise combinatorial definition of weak interaction remains a subject for future research on NP complete problems.

Bibliography

1. Applegate, D. and Cook, W. (1991). A computational study of the job-shop scheduling problem. *OSRA Jour Computing, 3(2)*, 149-156.

2. Atkinson, W. and Cunningham, J. (1991). Proving properties of a safety-critical system in FOREST. *Software Eng. J., 6(2)*, 41-50.

3. Cheng, T.C.E. and Sin, C.S.S. (1990). A state-of-the-art review of parallel-machine scheduling research. *Eur. J. Operational Res., 47*, 271-292.

4. Conroy, G.V. and Pulley, C. (1994). Logical methods in the formal verification of safety-critical software. *Proceedings 1993 IMA Conference "The Mathematics of Dependable Systems"*.

5. Cribbens, A.H. (1987). Solid State Interlocking (SSI): an integrated electronic signalling system for mainline railways. *Proceedings IEE, 134(3)*, pp. 148-158.

6. Cribbens, A.H. and Mitchell, I.H. (1992). The application of advanced computing techniques to the generation and checking of SSI data. *1991 Railway Engineers' Forum Meeting, also Proceedings IRSE, 1992*.

7. Davey, B.A. and Priestley, H.A. (1990). *Introduction to Lattices and Order: Chapter 11 Formal concept analysis*, Cambridge University Press.

8. Ganter, B. (1987). *Algorithmen zur formalen Begriffsanalyse*, Editors: B. Ganter, R. Wille and K.E. Wolff, B.I.-Wissenschaftsverlag Germany, 241-254.

9. Ganter, B. and Reuter, K. (1991). Finding all closed sets: a general approach. *Order, 8*, 283-290.

10. Ginzburg, A. (1972). *Algebraic Theory of Automata*, Third printing, Academic Press ACM Monograph, New York, USA.

11. Guenoche, A. and van Mechelen, I. (1992). Galois approach to the induction of concepts. *Categories and concepts: Theoretical views and inductive data analysis*, Editors: van Mechelen et al., Academic Press, London, pp. 287-308.

12. Hartmanis, J. and Stearns, R.E. (1966). *Algebraic Structure Theory of Sequential Machines*, Prentice-Hall, New Jersey, USA.

13. Hou, T-H. and Wang, H-P. (1991). Integration of a CAPP system and an FMS. *Computers and Ind. Engineering, 20(2)*, 231-242.

14. Husbands, P. (1993). An ecosystem model for integrated production planning. *Int. J. Computer Integrated Manufacturing, 6(1&2)*, 74-86.

15. Ingleby, M. and Mitchell, I.H. (1992). Proving safety of a railway signalling system incorporating geographic data. SAFECOMP 1992 Conference Proceedings on "IFAC", Editor: H.H.Frey, Pergamon Press, Zürich, Switzerland, pp. 129-134.

16. Ingleby, M. (1992). Formal validation of control data for signal interlocking. *AAR/BRR Conference Proceedings on "Reliability"*.

17. Iwata, K., Murotsu, Y., Oba, F. and Uemura, T. (1978). Optimization of selection of machine tools, loading sequence of parts and machining conditions in job-shop type machining systems. *Ann. CIRP, 27(1)*, 447-451.

18. Kelley, J.L. (1955). *General Topology*, van Nostrand, New York USA.

19. Khoshnevis, B. and Chen, Q-M. (1990). Integrating of process planning and scheduling functions. *J. Intelligent Manufacturing, 1*, 165-176.

20. Kleene, S.C. (1956). Representation of events in nerve nets and finite automata. *Automata Studies*, Princeton UP, New Jersey, USA, 3-41.

21. Krause, F.L. and Altmann, C. (1991). Integration of CAPP and scheduling. *Computer Applications in Production and Engineering: Integration Aspects*, Editors: G. Doumeingts, J. Browne and M. Tomljanovich, Elsevier Science Publishers/IFIP.

22. Lawler, E.L., Lenstra, L.K. and Rinooy Kan, A.H.G. (1982). Recent developments in deterministic sequencing and scheduling: A survey. *Deterministic and Stochastic Scheduling*, Editor: M.A.H. Dempster, D. Reidel Publishing, Amsterdam, The Netherlands, pp. 35-73.

23. Opitz, O., Lausen, B. and Klar, R. (1993). *Information and Classification: Concepts, Methods and Applications*, Springer-Verlag, Berlin, Germany.

24. Váncza, J. and Márkus (1991). Genetic algorithms in process planning. *Computers in Industry, 17*, 181-194.

25. van Laarhoven, P.J.M., Aarts, E.H.L. and Lenstra, J.K. (1992). Job shop scheduling by simulated annealing. *Operations Research, 40(1)*, 113-125.

26. Wille, R. (1984). Line diagrams of hierarchical concept systems. *International Classification, 11*, 77-86.

27. Wille, R. (1987). Bedeutungen von Begriffs-verbänden. Beiträge zur Begriffsanalyse, Editors: B. Ganter, R. Wille and K.E. Wolff, B.I.-Wissenschaftsverlag, Germany, pp. 161-211, .

28. Wille, R. (1989). Lattices in data analysis: how to draw them with a computer. *Algorithms and Order*, Editor: I. Rival, Kluwer publishing, Boston, USA.

29. Wille, R. (1992). Concept lattices and conceptual knowledge systems. *Computers and Mathematics with Applications*, *23*, 493-515.

A formal framework for fault-tolerant programs

Z. Liu* and M. Joseph**

**Department of Mathematics and Computer Science, University of Leicester and **Department of Computer Science, University of Warwick, Coventry*

Abstract This paper shows how general principles of program refinement can be used for the development of fault-tolerant programs. Let program P be represented by a predicate defining its initial condition and a set of atomic operations that carry out state transitions in a non-deterministic and interleaved manner. Let faults in the execution environment of P be represented by a set F of atomic operations. The effect of these faults can be modelled as a transformation of P into its F-affected version $\mathcal{F}(P, F)$. Then to prove that a *low level* program P_l is a F-tolerant implementation of a high level program P_h is to prove that $\mathcal{F}(P_l, F)$ *refines* P_h. Thus to prove the properties of fault-tolerant programs does not require the development of a new semanics along proof rules, languages, and tools.

1 Introduction

Consider the development of a concurrent program by a sequence of steps of refinement, from a requirement specification to an *executable* implementation. Each step in such a development constructs a *lower level* specification P_l from a *higher level* specification P_h of the program and it can be proved that P_l refines P_h, denoted $P_l \sqsubseteq P_h$. The refinement relation $P_l \sqsubseteq P_h$ is reflexive ($P \sqsubseteq P$ for any specification P), and transitive (if $P_{ll} \sqsubseteq P_l$ and $P_l \sqsubseteq P_h$, then $P_{ll} \sqsubseteq P_h$). Different frameworks and refinement calculi [1, 2, 3, 4] have been used to prove that one specification refines another.

The refinement steps described above assume that the implementation P_l is executed on a fault-free system, i.e. each operation in P_l is executed correctly, according to its defined semantics. However, concurrent programs may be run on systems which exhibits various failures, e.g. processors may fail, or channels may lose messages. Thus, if the execution of P_l suffers from the effects of such faults, its behaviour may not satisfy P_h even though $P_l \sqsubseteq P_h$ can be proved.

To deal with this problem, assume that the physical faults of a system are modelled as being caused by a set F of fault operations which perform state transformations in the same way as the ordinary program operations. The effect (i.e. interference) of the faults F on the execution of a program P can be modelled by a transformation $\mathcal{F}$ of P into an 'F-affected' version $\mathcal{F}(P, F)$. This F-affected version may not satisfy the specification of P. But a 'recovery' transformation $\mathcal{R}$ can sometimes be applied to introduce redundancy into the execution of P so that $\mathcal{F}(\mathcal{R}(P), F)$ satisfies the desired specification. When this is possible, $\mathcal{R}(P)$ can be called an F-tolerant version of P.

In [9] we presented a temporal semantic model of faults and fault-tolerance. In this, the F-affected version $\mathcal{F}(P, F)$ is simply the union of the operations of P and F and these are executed in an interleaved manner. Fault-tolerance is achieved if a specification P_l is an *F-tolerant implementation* of a specification (program) P_h and the F-affected version $\mathcal{F}(P_l, F)$ of P refines P_h.

This paper uses an action system model to formalize the specification of the fault-tolerant program framework presented in [9]. The Temporal Logic of Actions (TLA) [11] is used to reason about fault-tolerant programs.

2   Action system models

A program will be represented as an *action system*. The model used in this paper is simpler than the one defined in [10], but because of the use of control variables it does not lose very much in terms of expressiveness.

2.1   Definition of an action system

An action system is a pair $P = (I, AO)$ consisting of an *initial condition I* and a set AO of *atomic operations* on a finite set V of *state variables*. A *state s* of program P is a mapping from V to an associated set of values (i.e. the *value domain*) D. We use $s[x]$, for $x \in V$, to denote the value of x in state s.

The initial condition I is a first order predicate with free variables in V. It defines the permitted initial states from which execution of P can start.

Each atomic operation $\tau \in AO$ is represented by a guarded nondeterministic multiple assignment $g \rightarrow X := X_0.Q$, where X is a vector of some state variables in V, X_0 is a vector of logical variables, and Q is a first order predicate over state variables and logical variables. When the guard g of τ is *true* (i.e. *τ is enabled*) in a state s, τ can be executed and will change state s into a state t such that

- the variables in X are assigned the values X_0 and the other variables remain unchanged

$$t[y] = \begin{cases} X_0(i) & \text{if } y \text{ is the } i\text{th element } X(i) \text{ of } X \\ s[y] & \text{otherwise} \end{cases}$$

- the predicate Q is satisfied by t.

An atomic operation may introduce infinite nondeterminism. This expressive power is needed to represent a fault by an atomic operation which may damage the program state by setting a variable with any value but the *correct one*.

A deterministic assignment is a special case of a nondeterministic assignment. We use $g \to X := E$ for

$$g \to X := X_0.(\bigwedge_{i=1}^{|X_0|} X_0(i) = E(i))$$

where $|X_0|$ is the length of the vector X_0, and each $E(i)$ is an expression which is the ith element of vector E. The guard g of an operation may be omitted when it is the constant *true*.

Every operation is *atomic*: once started, its execution cannot be interrupted and intermediate states during the execution cannot be observed by other operations. Because of the atomicity of operations, a *behaviour* (or computation) of program P can be represented as an infinite sequence $\sigma = s_0, s_1, s_2, \ldots$ of states, where s_0 is the *initial state* which satisfies the initial condition I, and each s_{i+1}, $i \geq 0$, is obtained from s_i by either a *stuttering step*, i.e. $s_{i+1} = s_i$, or by executing an operation τ of program P.

The set of all the behaviours of a program is *stuttering closed*: if an infinite state sequence σ is a behaviour of the program, then so is any behaviour obtained from σ by adding or deleting a finite number of stuttering steps. A behaviour is *terminating* if it has an infinite number of stuttering steps.

Example 1. *Consider a simple processor-memory interface. The processor issues* read *and* write *operations that are executed by the memory. Such an interface consists of two registers, represented by the following two state variables.*

op: Set by the processor to indicate the desired operation, and reset by the memory after executing the operation. Its value space is

$$\{ready, read, write\}$$

val: Set by the processor to indicate the value to be written by a write, and set by the memory to return the result of a read. Assume that its value space is the integers $\mathbf{Z}$.

To describe the interface as an action system, we introduce an (internal) variable *mem* with value space $\mathbf{Z}$ to denote the contents of the memory.

Let

$$I \quad \triangleq \quad (op = ready)$$

$$R_p \quad \triangleq \quad (op = ready) \rightarrow op := read$$

$$W_p \quad \triangleq \quad (op = ready) \rightarrow (op, val) := (w, v).(w = write \wedge v \in \mathbf{Z})$$

$$R_m \quad \triangleq \quad (op = read) \rightarrow (op, val) := (ready, mem)$$

$$W_m \quad \triangleq \quad (op = write) \rightarrow (op, mem) := (ready, val)$$

$$AO \quad \triangleq \quad \{R_p, W_p, R_m, W_m\}$$

The interface can be described as the program

$$InterFace \triangleq (I, AO)$$

2.2 Concurrency

The parallel execution of a program $P = (I, AO)$ can be modelled by partitioning its *operations* AO into a set $Proc = \{p_1, \ldots, p_k\}$ of processes, as in Back's *concurrent action system* [10]. Let a *shared state variable* be one that is used by two or more operations in different processes, and a *private state variable* be used only by the operations in one process. Then two operations of P can be executed in parallel if and only if they are not in the same process and do not share variables. A concurrent program P with a set $Proc$ of k processes can be written as

$$P = (I, p_1 \cup \ldots \cup p_k)$$

In a concurrent program, processes communicate by executing operations which use share variables.

A sequential program is a special case of a parallel program where all the operations are assigned to the same process. Another special case is where each operation is assigned to a separate process. The program is then executed with maximal parallelism.

Because operations are executed atomically, a parallel computation can be modelled by a sequential computation with interleaved execution of operations. Hence, the set of possible executions of an action system will be the same for a concurrent or a distributed action system and for a sequential action system. This allows us to separate the logical behaviour of processes from implementation issues. The same principle is also used in [11, 12].

Each action system is assumed to be a closed system, i.e. the environment is included in the 'system'. Operations AO therefore consist of those carried out by the program, and of those carried out by the environment.

Using a closed system makes it possible for a fault-tolerant system to guarantee that recovery is always possible by restoring processes to a consistent state [5,7,8,9]. If a program interacts with an external environment (as in an open system), recovery of the program must be accompanied by the restoration to a consistent state of the whole system. For example, it may be necessary that after recovery, communications with external devices not be repeated, or that the program is restored to a possible future state (i.e. using forward recovery) that is consistent with the states of the external devices.

3 Reasoning about action systems in TLA

To reason formally about the properties of an action system, we shall use TLA. Models of TLA are defined as infinite sequences of states, $\sigma = s_0, s_1, \ldots$, where each s_i is a state over a set of state variables. A property of program P is represented by a formula φ which defines a set of behaviours of P.

An *action formula* $\mathcal{A}$ is a boolean-valued expression which specifies the relation between the values of variables before the execution of an operation and the values of 'primed' variables after the execution of the operation. It is interpreted over pairs of states $\langle s, s' \rangle \models \mathcal{A}$ such that $\mathcal{A}$ holds when the variables V in $\mathcal{A}$ are assigned values according to state s, and the primed variables $V' = \{x' \mid x \in V\}$ in $\mathcal{A}$ are assigned values according to state s'. $\langle s, s' \rangle$ is called an $\mathcal{A}$-step if $\langle s, s' \rangle \models \mathcal{A}$. Formulas in TLA are constructed from action formulas in the following way.

An action formula $\mathcal{A}$ is a formula. $\mathcal{A}$ is satisfied by a sequence $\sigma = s_0, s_1, s_2, \ldots$, of program P, denoted $\sigma \models \mathcal{A}$, if $\langle s_0, s_1 \rangle \models \mathcal{A}$.

A first order state predicate ψ is a particular action formula which does not have primed variables; σ satisfies ψ if the initial state s_0 of σ satisfies ψ, i.e. $s_0 \models \psi$.

If φ is a formula, so is $\Box\varphi$ (always φ). σ satisfies $\Box\varphi$ if all the suffixes of σ satisfy φ. The formula $\Diamond\varphi$ (eventually φ) is defined to as $\neg\Box\neg\varphi$, and this is satisfied by σ if there is a suffix of σ which satisfies φ. Formulas can be composed using the first order connectives $\neg$, $\wedge$, $\vee$ and $\Rightarrow$ with their standard semantics. Quantification (i.e. $\exists x\varphi$, $\forall x\varphi$) is possible over logical variables, whose values are fixed over states, and over state variables, whose values can change from state to state.

For a vector U state variables, let

$$unch(U) \triangleq \bigwedge_{x \in U} (x' = x)$$

An $unch(U)$-step $\langle s, s' \rangle$ is also called an U-stuttering step which leaves the variables in U unchanged.

For an action formula $\mathcal{A}$ and a vector of U of state variables, let

$$[\mathcal{A}]_U \triangleq \mathcal{A} \vee unch(U)$$

3.1   Transforming action systems to TLA formulas

The relationship between action systems and TLA can be defined in a relatively straightforward way. A syntactic operation $\tau : g \rightarrow X := X_0.Q$ in a action system P corresponds to an equivalence class of action formulas. For each action formula $\mathcal{A}$ in this class and each pair $\langle s, s' \rangle$ of states over the state variables V of P, execution of τ in state s terminates in state s' iff $\langle s, s' \rangle$ is an $\mathcal{A}$-step.

The operation τ can be translated into an action formula $\Gamma(\tau)$ which is a representative of the equivalence class of action formulas corresponding to τ.

$$\Gamma(\tau) \triangleq g \wedge Q[X(i)'/X_0(i)] \wedge unchd(V \setminus X)$$

where $X(i)'$ is the primed counterpart of the ith variable in X, and $V \setminus X$ is the complement of the variables X over the program variables V of P.

Action τ is enabled only when g holds. When τ is enabled and executed, the program state is changed so that the new values of the variables X are related to the old values in the way defined by $Q[X(i)'/X_0(i)]$, and the values of any variable outside X remain unchanged.

Given an action system (program) $P = (I, AO)$, let

$$\mathcal{N}_P \triangleq \bigvee_{\tau \in AO} \Gamma(\tau)$$

This is the *state-transition relation* for the atomic operations of P.

The *internal safety specification* of P is expressed by the formula

$$\Pi(P) \triangleq I \wedge \Box[\mathcal{N}_P]_V$$

This formula specifies all the possible sequences of values that may be taken by the state variables, including the *internal* variables.

Using existential quantification to hide the internal variables, $X = (x_1, \ldots, x_n) \subseteq V$, the (*external*) *safety specification* of P is given as

$$\Phi(P) \triangleq \exists X : \Pi(P)$$

A behaviour satisfies $\Phi(P)$ iff there is a sequence of values that can be assigned to x_i, $i = 1, \ldots, n$, that makes $\Pi(P)$ true. (A complete definition can be found in [11].)

Example 1 (continued). *For the processor-memory interface, let*

$$
\begin{aligned}
\Gamma(R_p) &\triangleq (op = ready) \wedge (op' = read) \wedge unchd(mem, val) \\
\Gamma(W_p) &\triangleq (op = ready) \wedge (op' = write) \wedge (val' \in \mathbf{Z}) \wedge unchd(mem) \\
\Gamma(R_m) &\triangleq (op = read) \wedge (op' = ready) \wedge (val' = mem) \wedge unchd(mem) \\
\Gamma(W_m) &\triangleq (op = write) \wedge (op' = ready) \wedge (mem' = val) \wedge unchd(val)
\end{aligned}
$$

Then

$$
\Pi(InterFace) = (op = ready) \wedge \Box[\Gamma(R_p) \vee \Gamma(W_p) \vee \Gamma(R_m) \vee \Gamma(W_m)]_V
$$

where $V \triangleq \{op, val, mem\}$.

From the users' point of view, mem in this specification is an internal variable. Thus, we have

$$
\Phi(InterFace) \triangleq \exists mem : (op = ready) \wedge \Box[\Gamma(R_p) \vee \Gamma(W_p) \vee \Gamma(R_m) \vee \Gamma(W_m)]_V
$$

Note that in the syntax of action systems, there is no equivalent to such internal variables. It would be possible to rewrite the specification $P = (I, AO)$ of program P as $P = \mathbf{var}\, X : (I, AO)$ which makes explicit the declaration of *internal variables*. However, we shall just leave the treatment of internal variable declaration informal.

3.2   Imposing fairness properties

Formulas $\Pi(P)$ and $\Phi(P)$ are safety properties, i.e. *they are satisfied by an infinite behaviour iff they are satisfied by every finite initial portion of the behaviour.* Safety properties allow behaviours in which a system performs correctly for a while and then leaves the values of all variables unchanged. Such behaviours are undesirable in distributed systems and they can be ruled out by adding fairness properties.

For an action formula $\mathcal{A}$, let the state predicate *Enabled $\mathcal{A}$* be true for state s if there exists a state t such that (s, t) is an $\mathcal{A}$-step. For a vector U of variables in V, let

$$
\Diamond\langle\mathcal{A}\rangle_U \triangleq \neg\Box[\neg\mathcal{A}]_U
$$

defines $\langle\mathcal{A}\rangle_U$ as the action formula $\mathcal{A} \wedge \neg unchd(U)$. The *weak* and *strong* fairness conditions for an action formula $\mathcal{A}$ have the respective forms:

$$WF_U(\mathcal{A}) \triangleq (\Box\Diamond\langle\mathcal{A}\rangle_U) \lor (\Box\Diamond\neg Enabled\langle\mathcal{A}\rangle_U)$$

$$SF_U(\mathcal{A}) \triangleq (\Box\Diamond\langle[\mathcal{A}\rangle_U) \lor (\Diamond\Box\neg Enabled\langle\mathcal{A}\rangle_U)$$

The weak fairness condition $WF_U(\mathcal{A})$ says that from any point in a behaviour, either $\langle A\rangle_U$ will be disabled infinitely often or an $\langle A\rangle_U$-step will eventually occur. The strong fairness condition $WF_U(\mathcal{A})$ says that from any point in a behaviour, either $\langle A\rangle_U$ is enabled only finitely often or a $\langle A\rangle_U$-step will eventually occur.

The safety specifications $\Pi(P)$ and $\Phi(P)$ are usually strengthened by conjoining them with one or more fairness properties. The *canonical forms* of the internal and external specifications of a P are respectively

$$\Psi(P) = I \land \Box[\mathcal{N}_P]_V \land \mathcal{L}$$

$$S(P) \triangleq \exists X : (I \land \Box[\mathcal{N}_P]_V \land \mathcal{L})$$

where $X = \{x_1, \ldots, x_n\}$ is the tuple of the internal variables of P, the action formula $\mathcal{N}_P$ defines the next-state relation of P, V is the set of variables of P and $\mathcal{L}$ is a conjunction of the fairness properties of action formulas which defines the fairness requirement of P.

Example 1 (continued). *For the processor-memory interface*

$$
\begin{aligned}
Enabled\langle\Gamma(R_m)\rangle_V &= op = read \\
Enabled\langle\Gamma(W_m)\rangle_V &= op = write
\end{aligned}
$$

$$
\begin{aligned}
WF_V(\Gamma(R_m)) &= (\Box\Diamond\langle\Gamma(R_m)\rangle) \lor (\Box\Diamond(op \neq read)) \\
SF_V(\Gamma(R_m)) &= (\Box\Diamond\langle\Gamma(R_m)\rangle) \lor (\Diamond\Box(op \neq read))
\end{aligned}
$$

Similarly, we can obtain $WF_V(\Gamma(W_m))$ and $SF_V(\Gamma(W_m))$.

We assume weak fairness conditions for $\Gamma(R_m)$ and $\Gamma(W_m)$,

$$\mathcal{L} \triangleq WF_V(\Gamma(R_m)) \land WF_V(\Gamma(W_m))$$

Given an action system P with specification $S(P)$, let a property of P be described by the formula $\mathcal{P}$. To prove that P satisfies property $\mathcal{P}$ is to prove in TLA the validity of the implication

$$S(P) \Rightarrow \mathcal{P}$$

For example, a *liveness* (progress) property often has the form $Q_1 \longmapsto Q_2$ (read as Q_1 *leads-to* Q_2), and this can be defined as

$$Q \longmapsto Q_2 \triangleq \Box(Q_1 \Rightarrow \Diamond Q_2)$$

To prove that P satisfies $Q_1 \longmapsto Q_2$ is just to prove

$$S(P) \Rightarrow (Q_1 \longmapsto Q_2)$$

3.3 Refinement mapping

The relation $P_1 \sqsubseteq P_2$ characterizes *refinement*, i.e. that program P_1 correctly implements P_2. Let

$$\Phi(P_1) = \exists X : I_1 \wedge \Box[\mathcal{N}_{P_1}]_{V_1} \qquad \Phi(P_2) = \exists Y : I_2 \wedge \Box[\mathcal{N}_{P_2}]_{V_2}$$

be safety specifications of P_1 and P_2 respectively, where

$$X = \{x_1, \ldots, x_n\} \qquad Y = \{y_1, \ldots, y_m\}$$

Then the *safety refinement relation* is formalized as

$$P_1 \sqsubseteq_s P_2 \quad \text{iff} \quad \Phi(P_1) \Rightarrow \Phi(P_2)$$

This relation can be strengthened by taking liveness properties into account. Let

$$S(P_1) = \exists X : I_1 \wedge \Box[\mathcal{N}_{P_1}]_{V_1} \wedge \mathcal{L}_1$$

$$S(P_2) = \exists Y : I_2 \wedge \Box[\mathcal{N}_{P_2}]_{V_2} \wedge \mathcal{L}_2$$

be the canonical specifications of P_1 and P_2 respectively. The refinement P_1 of P_2 taking into account the fairness conditions $\mathcal{L}_1$ and $\mathcal{L}_2$ is

$$P_1 \sqsubseteq P_2 \quad \text{iff} \quad S(P_1) \Rightarrow S(P_2)$$

Clearly, the safety refinement is a special case of the refinement with fairness conditions when both $\mathcal{L}_1$ and $\mathcal{L}_2$ are *true*.

The refinement relation $P_1 \sqsubseteq P_2$ can be proved as the implication $S(P_1) \Rightarrow S(P_2)$. And to prove this implication, we must define state functions $\overline{y}_1, \ldots, \overline{y}_m$ in terms of the variables V_1 and prove the implication

$$\Psi(P_1) \Rightarrow \overline{\Psi(P_2)}$$

where $\overline{\Psi(P_2)}$ is obtained from $\Psi(P_2)$ by substituting $\overline{y}_i$ for all the free occurrences of y_i in $\Psi(P_2)$, for $i = 1, \ldots, m$. The collection of state functions $\overline{y}_1, \ldots, \overline{y}_m$ is called a *refinement mapping*[1].

4 Faults and their effects

Let $P = (I, AO)$ be a program with the safety specification

$$\Phi(P) = \exists X : I \wedge \Box[\mathcal{N}_P]_V$$

---

[1] The validity of the implication $S(P_1) \Rightarrow S(P_2)$ does not imply the existence of a refinement mapping. In general, refinement mappings can be found if we modify the specifications by adding dummy variables [2].

Informally speaking, a *physical fault* in the execution of P causes a transition from a valid state of P into an *error* state. Continuation of the execution of P from such an error state may leads to a *failure* state which deviates from the specification of P.

Thus in general, a physical fault can be modelled as an effect of an atomic *fault-operation*; this can be translated to a action formula by Γ which is presented in Section 3.1.

For example, an arbitrary malicious fault may set the variables of P to arbitrary values; a crash in a processor may cause the variables to become unavailable; and a fault may cause the loss of a message from a channel. Physical faults can be described by a set of atomic operations F , called a *fault-environment*, which interfere with the execution of P by possibly changing the values of variables in V. F can be specified by the action formula $\mathcal{N}_F$

$$\mathcal{N}_F = \bigvee_{\tau \in F} \Gamma(\tau)$$

Executing $P = (I, AO)$ on a system with a fault-environment F is simulated by interleaving the execution of the operations of P and F. Therefore, interference by F on the execution of P can be defined as a transformation $\mathcal{F}$

$$\mathcal{F}(P, F) \triangleq (I, AO \cup F)$$

The internal and external behaviours of P being executed on a system with faults F are then specified respectively by

$$\Pi(\mathcal{F}(P, F)) = I \wedge \Box[\mathcal{N}_P \vee \mathcal{N}_F]_V$$

and

$$\Phi(\mathcal{F}(P, F)) = \exists X : I \wedge \Box[\mathcal{N}_P \vee \mathcal{N}_F]_V$$

The fault-prone properties of P under F can be reasoned about in terms of the properties of $\mathcal{F}(P, F)$. We call $\mathcal{F}(P, F)$ the *F-affected version* of P, and a behaviour of $\mathcal{F}(P, F)$ an *F-affected behaviour* of P.

Example 1 (continued). *For the processor-memory interface, assume that the memory is faulty and its value may be corrupted. Such a fault can be represented by the atomic operation*

$$fault \triangleq true \rightarrow mem := mem_0.(mem_0 \neq mem)$$

whose action formula is

$$\Gamma(fault) = (mem' \neq mem) \wedge unchd(val, op)$$

Let

$$F \triangleq \{fault\}$$

The F-affected version of InterFace is

$$\mathcal{F}(InterFace, F) = ((op = ready), \{R_p, W_p, R_m, W_m, fault\})$$

Thus,

$$\Pi(\mathcal{F}(InterFace, F)) = \quad (op = ready) \wedge \Box[\Gamma(R_p)$$
$$\Gamma(W_p) \vee \Gamma(R_m) \vee \Gamma(W_m) \vee \Gamma(fault)]_V$$

where

$$V \triangleq \{op, val, mem\}$$

5 Fault-tolerance

Given a program $P = (I, AO)$ with the specification

$$S(P) = \exists X : (I \wedge \Box[\mathcal{N}_P]_V \wedge \mathcal{L})$$

and a property described by a formula φ, P is said to *implement* (or *satisfy*) φ if $S(P) \Rightarrow \varphi$.

Such an implementation assumes that the hardware is fault-free, i.e. that each operation of P is executed according to its semantics. However, if P is run on a system subject to hardware faults, errors may occur when the program is executed and lead to failures which violate the specification S. That is, when a set F of faults interferes with the execution of P, the F-affected behaviours may not satisfy φ, and $\mathcal{F}(P, F)$ is then not an implementation of φ.

For P to be fault-tolerant, *correcting operations* must be carried out to prevent an error from leading to a failure. P is called an F-tolerant implementation of φ, if $\mathcal{F}(P, F)$ is an implementation of φ:

$$S(\mathcal{F}(P, F)) \Rightarrow \varphi$$

This means that the behaviours of P comply with the specification φ despite the presence of faults F, i.e. the implementation P of φ tolerates the faults described by F.

When such a property φ is a canonical specification of a program P_h

$$S(P_h) = \exists Y : I_h \wedge \Box[\mathcal{N}_{P_h}]_U \wedge \mathcal{L}_h$$

P_l is called an *F-tolerant refinement* of P_h, denoted $P_l \sqsubseteq_F P_h$, if P_l is a F-tolerant implementation of $S(P_h)$.

The F-tolerant refinement relation $\sqsubseteq_F$ is stronger than the ordinary refinement relation; i.e. if P_l is a F-tolerant refinement of P_h, then P_l is a refinement of P_h but the converse is not necesarily true. F-tolerant refinement is generally not reflexive, i.e. $P \sqsubseteq_F P$ does not generally hold.

However, it is transitive: if $\mathcal{F}(P_{ll}, F_1) \sqsubseteq P_l$ and $\mathcal{F}(P_l, F_2) \sqsubseteq P_h$, then $\mathcal{F}(P_{ll}, F_1) \sqsubseteq P_h$. This allows a stepwise-style development of a fault-tolerant program. Faults (or their representations) considered at a lower level step in the development may differ from those in the previous step.

Note that this definition of fault-tolerance for a program is with respect to the specification of the program. This distinguishes it from some other definitions, e.g. [13], where P is F-tolerant only if $P \sqsubseteq_F P$.

Example 1 (continued). *Assume we must prove that the fault-free memory InterFace can be implemented using three faulty memories whose values may be corrupted, under the condition that at any time at most one can be corrupted.*

Let mem_1, mem_2 and mem_3 denote the current values of the three memories respectively. Each memory is faulty and its value mem_i may be corrupted by $fault_i$, $i = 1, 2, 3$.

Let f_i be a variable with value space $\{0, 1\}$ and assume that mem_i has been corrupted if $f_i = 1$. Each $fault_i$ can be specified as follows.

$$\begin{aligned}
\Gamma(fault_1) \; &\triangleq \; &&(f_2 = 0 \wedge f_3 = 0) \wedge (mem_1' \neq mem_1) \wedge (f_1' = 1) \\
&&\wedge \;\; &unchd(val, mem_2, mem_3, f_2, f_3)
\end{aligned}$$

$$\begin{aligned}
\Gamma(fault_2) \; &\triangleq \; &&(f_1 = 0 \wedge f_3 = 0) \wedge (mem_2' \neq mem_2) \wedge (f_2' = 1) \\
&&\wedge \;\; &unchd(val, mem_1, mem_3, f_1, f_3)
\end{aligned}$$

$$\begin{aligned}
\Gamma(fault_3) \; &\triangleq \; &&(f_1 = 0 \wedge f_2 = 0) \wedge (mem_3' \neq mem_3) \wedge (f_3' = 1) \\
&&\wedge \;\; &unchd(val, mem_1, mem_2, f_1, f_2)
\end{aligned}$$

$$F \quad \triangleq \quad \{fault_1, fault_2, fault_3\}$$

$$\mathcal{N}_F \quad \triangleq \quad \Gamma(fault_1) \vee \Gamma(fault_2) \vee \Gamma(fault_3)$$

Each $fault_i$, for $i = 1, 2, 3$ can be represented as the operation

$$fault_i \triangleq$$

$$(f_{i \oplus 1} = 0 \wedge f_{i \ominus 1} = 0) \rightarrow (mem_i, f_i) := (mem, f).(mem \neq mem_i \wedge f = 1)$$

The specification of a program P_l, which is the F-tolerant refinement of InterFace, can be obtained by specifying its variables, initial condition, operations and fairness conditions.

The variables and initial condition are as follows.

$$V_l \triangleq \{op, val, mem_1, mem_2, mem_3, f_1, f_2, f_3\}$$

$$I_l \triangleq (op = ready) \wedge (f_1 = f_2 = f_3 = 0)$$
$$\wedge \ (mem_1 = mem_2 = mem_3)$$

To specify the operations of P_l, we first define the following auxiliary function.

$$vote(x, y, z) \triangleq \left\{ \begin{array}{ll} x & \text{if } x = y \text{ or } x = z \\ y & \text{if } x \neq y \text{ and } x \neq z \end{array} \right.$$

The specification of the rest of the program is:

$$\Gamma(R_p^l) \triangleq (op = ready) \wedge (op' = read)$$
$$\wedge \ unchd(val, mem_1, mem_2, mem_3, f_1, f_2, f_3)$$

$$\Gamma(W_p^l) \triangleq (op = ready) \wedge (op' = write) \wedge (val' \in \mathbf{Z})$$
$$\wedge \ unchd(mem_1, mem_2, mem_3, f_1, f_2, f_3)$$

$$\Gamma(R_m^l) \triangleq (op = read) \wedge (op' = ready)$$
$$\wedge \ (val' = vote(mem_1, mem_2, mem_3))$$
$$\wedge \ unchd(mem_1, mem_2, mem_3, f_1, f_2, f_3)$$

$$\Gamma(W_m^l) \triangleq (op = write) \wedge (op' = ready) \wedge (\bigwedge_{i=1}^{3}(mem_i' = val))$$
$$\wedge \ (\bigwedge_{i=1}^{3}(f_i' = 0)) \wedge unchd(val)$$

$$\mathcal{N}_{P_l} \triangleq \Gamma(R_p^l) \vee \Gamma(W_p^l) \vee \Gamma(R_m^l) \vee \Gamma(W_m^l)$$

$$L_l \triangleq WF(\Gamma(R_m^l)) \cap WF(\Gamma(W_m^l))$$

$$\Psi(P_l) \triangleq I_l \wedge \Box[\mathcal{N}_{P_l}]_{V_l} \wedge L_l$$

$$S(P_l) \triangleq \exists mem_1, mem_2, mem_3, f_1, f_2, f_3 : \Psi(P_l)$$

We define the following operations:

$$R_p^l \triangleq (op = ready) \rightarrow op := read$$
$$W_p^l \triangleq (op = ready) \rightarrow (op, val) := (w, v).(w = write \wedge v \in \mathbf{Z})$$
$$R_m^l \triangleq (op = read) \rightarrow (op, val) := (ready, vote(mem_1, mem_2, mem_3))$$
$$W_m^l \triangleq (op = write) \rightarrow (op, (mem_i), (f_i)) := (ready, (val), (0))$$

where (mem_i) is an abbreviation for vector (mem_1, mem_2, mem_3), (f_i) for vector (f_1, f_2, f_3), (val) for vector (val, val, val) and (0) for vector $(0, 0, 0)$. Let P_l be defined as

$$P_l \triangleq (I_l, \{R_p^l, W_p^l, R_m^l, W_m^l\})$$

The action formulas of R_p^l, W_p^l, R_m^l and W_m^l are respectively $\Gamma(R_m^l)$, $\Gamma(W_p^l)$, $\Gamma(R_m^l)$ and $\Gamma(W_m^l)$.

And we can formally prove the fault assumption that at any time at most one memory can be corrupted by faults,

$$\Psi(\mathcal{F}(P_l, F)) \Rightarrow \Box(f_1 + f_2 + f_3 \leq 1)$$

where $\Psi(\mathcal{F}(P_l, F)) = I_l \wedge \Box[\mathcal{N}_{P_l} \vee \mathcal{N}_F]_{V_l} \wedge L_l$

We can also prove that P_l is a F-tolerant refinement of the fault-free InterFace. To do so, notice that the specification of the F-affected version of P_l is

$$S(\mathcal{F}(P_l, F)) = \exists mem_1, mem_2, mem_3, f_1, f_2, f_3 : I_l \wedge \Box[\mathcal{N}_{P_l} \vee \mathcal{N}_F]_{V_l} \wedge L_l$$

Define the following mappings from V_l to V

$$\overline{op} = op$$

$$\overline{val} = val$$

$$\overline{mem} = vote(mem_1, mem_2, mem_3)$$

Using the proof rules of TLA, we then can easily prove the implication

$$S(\mathcal{F}(P_l, F)) \Rightarrow S(InterFace)$$

which shows that P_l is an F-tolerant refinement of InterFace.

It will be noticed that the fault assumption, that at any time at most one memory can be corrupted by faults, is encoded into the specification of the fault operations F and hence into the fault-affected version $\mathcal{F}(P_l, F)$:

$$\Psi(\mathcal{F}(P_l, F)) \Rightarrow \Box(f_1 + f_2 + f_3 \leq 1)$$

We can also treat the specification of fault assumptions (or the *global fault assumptions* of [14]) separately from the specifications of the state transitions carried out by faults (the *local fault assumptions*, cf. [14]). Let the set F_1 faults consists of $\{fault_{11}, fault_{12}, fault_{13}\}$ specified as

$$
\begin{aligned}
\Gamma(fault_{11}) \;\triangleq\; & (mem_1' \neq mem_1) \wedge (f_1' = 1) \\
\wedge\; & unchd(val, mem_2, mem_3, f_2, f_3)
\end{aligned}
$$

$$
\begin{aligned}
\Gamma(fault_{12}) \;\triangleq\; & (mem_2' \neq mem_2) \wedge (f_2' = 1) \\
\wedge\; & unchd(val, mem_1, mem_3, f_1, f_3)
\end{aligned}
$$

$$
\begin{aligned}
\Gamma(fault_{13}) \;\triangleq\; & (mem_3' \neq mem_3) \wedge (f_3' = 1) \\
\wedge\; & unchd(val, mem_1, mem_2, f_1, f_2)
\end{aligned}
$$

In this case, the $fault_{1i}$ for $i = 1, 2, 3$ can be simply represented as the action

$$
fault_{1i} \triangleq (mem_i, f_i) := (mem, f).(mem \neq mem_i \wedge f = 1)
$$

However with this, the implication $S(\mathcal{F}(P_l, F_1)) \Rightarrow S(InterFace)$ does not hold any more. Let

$$
FaultAssum \triangleq \Box(f_1 + f_2 + f_3 \leq 1)
$$

be the specification of the fault assumption, and let this be imposed on the specification $S(\mathcal{F}(P_l, F_1))$.

$$
\begin{aligned}
Spec(\mathcal{F}(P_l, F_1)) \triangleq\; & \exists mem_1, mem_2, mem_3, f_1, f_2, f_3 : I_l \wedge \Box[\mathcal{N}_{P_l} \vee \mathcal{N}_{F_1}]_{V_l} \\
& \wedge\, FaultAssum \wedge L_l
\end{aligned}
$$

It is easy to prove

$$
\Box[\mathcal{N}_{P_l} \vee \mathcal{N}_{F_1}]_U \wedge FaultAssum \Leftrightarrow \Box[\mathcal{N}_{P_l} \vee \mathcal{N}_F]_U
$$

Thus, we have

$$
Spec(\mathcal{F}(P_l, F_1)) \Leftrightarrow \Psi(\mathcal{F}(P_l, F))
$$

and hence

$$
Spec(\mathcal{F}(P_l, F_1)) \Rightarrow S(InterFace)
$$

This says that P_l can tolerate the faults F_1 of the assumption $FaultAssum$.

In general, the specification of the behaviours of the F-affected version $\mathcal{F}(P_l, F)$ of a program P_l can be of the form

$$
Spec(\mathcal{F}(P_l, F)) = \exists X : I_l \wedge \Box[\mathcal{N}_{P_l} \vee \mathcal{N}_F]_{V_l} \wedge FaultAssum \wedge L_l
$$

where I_l is the initial condition, $\mathcal{N}_l$ is a state transition relation which specifies the allowed state transitions, $\mathcal{N}_F$ is a state-transition relation which specifies the state transitions caused out by faults, $FaultAssum$ is a safety property which prevents certain transitions from taking place, L_l is a liveness property and V_l is the set of variables of P_l. P_l is a F-tolerant refinement of P_h under fault assumption $FaultAssum$ if the following implication can be proved:

$$Spec(\mathcal{F}(P_l, F)) \Rightarrow S(P_h)$$

Since $\Box[\mathcal{N}_{P_l} \vee \mathcal{N}_F]_{V_l}$ is a safety property and $FaultAssum$ is a safety properties ruling out certain transitions during the execution, the conjunction

$$\Box[\mathcal{N}_{P_l} \vee \mathcal{N}_F]_{V_l} \wedge FaultAssum$$

is again a safety property and hence can be transformed into a formula of the form $\Box[\mathcal{N}]_{V_l}$ where $\mathcal{N}$ is a new state-transition relation.

The separation of local and global assumptions about faults makes it easier to use the form $Spec(\mathcal{F}(P_l, F))$ for specifying the F-affected behaviours of program P_l. To prove that P_l is an F-tolerant refinement of P_h, it is helpful to transform this to the canonical form $\exists X : I_l \wedge \Box[\mathcal{N}]_U \wedge L_l$ so that the proof rule for refinement given in [2, 15, 16] can be used.

6   Summary

Using simple action system models and TLA, this paper formalizes the basic technique semantically described in [9] for the specification and verification of fault-tolerant programs. A requirement specification S is generally refined into a canonical specification $S(P_h)$ which can be encoded as an action system P_h. Analysis of the possible failures in the system (perhaps in comsultation with system engineers) provides the specification F (and/or $\mathcal{N}_F$) of the fault-environment. P_h is then refined into a F-tolerant implementation P_l. This was illustrated using the example of the processor-memory interface. Application of this transformational approach to the specification and verification of different fault-tolerant mechanisms can be found in [5, 6, 9].

Since behaviours specified in TLA consist only of state sequences, no distinction can be made about whether state changes are caused by the program operations or fault operations. The use of action systems allows such a distinction to be made.

Acknowledgements

We are grateful to the referee for suggesting improvements to the paper. The work was supported by research grant GB/F57960 when Zhiming Liu was a research fellow in the Department of Computer Science of the University of Warwick.

Bibliography

1. Morris, J.M. (1987). A theoretical basis for stepwise refinement and the programming calculus. *Science of Computer Programming, 9*, 287-306.

2. Abadi, M. and Lamport, L. (1988). The existence of refinement mapping. *Proceedings of the Third IEEE Sympoium on "Logic and Computer Science".*

3. Back, R.J.R. (1989). Refinement calculus, Part II: Parallel and reactive programs. *Technical Report, 93*, Abo Akademi.

4. Morgan, C. (1990). *Programming From Specifications*, Prentice Hall International Series in Computer Science.

5. Liu, Z. (1991). *Fault-Tolerant Programming By Transformations*, PhD thesis, Department of Computer Science, University of Warwick.

6. Liu, Z. and Joseph, M. (1992). Transformation of programs for fault tolerance. *Formal Aspects of Computing, 4(5)*, 442-469.

7. Peled, D. and Joseph, M. (1993). A compositional approach to fault-tolerance using formula transformation. *Lecture Notes in Computer Science, Vol. 694*, Springer-Verlag.

8. Peled, D. and Joseph, M. A compositional framework for fault-tolerance by specification transformation. *Theoretical Computer Science*, (to appear).

9. Liu, Z. and Joseph, M. (1993). Specifying and verifying of recovery in asynchronous communicating systems. *Formal Techniques in Real-Time and Fault Tolerant Systems*, Editor: J. Vytopil, Kluwer Academic Publishers, 137-166, (to appear).

10. Back, R.J.R. (1988). Refining atomicity in parallel algorithms. *Technical Report, 57*, Abo Akademi.

11. Lamport, L. (1991). The temporal logic of actions. *Technical Report, 79*, Digital SRC, California.

12. Chandy, K.M. and Misra, J. (1988). *Parallel Program Design: A Foundation*, Addison-Wesley Publishing Company.

13. Weber, D.G. (1993). Fault tolerance as self-simlarity. *Formal Techniques in Real-Time and Fault Tolerant Systems*, Editor: J. Vytopil-pages, Kluwer Academic Publishers, 33-49, (to appear).

14. Nordahl, J. (1992). *Specification and Design of Dependable Communicating Systems*, PhD thesis, Department of Computer Science, Technical University of Denmark.

15. Abadi, M. and Lamport, L. (1990). Composing specifications. *Technical Report*, *66*, Digital SRC, California.

16. Diepstraten, E. and Kuiper, R. (1990). Abadi & Lamport: towards a proof theory for stuttering, dense domains and refinement mappings. *Lecture Notes in Computer Science*, *430*: *Proceedings of the REX Workshop on "Stepwise Refinement of Distributed Systems, Models, Formalisms Correctness"*, pp. 208-238, Springer-Verlag.

Composition and refinement of probabilistic real-time systems

Z. Liu*, J. Nordahl and E.V. Sørensen****

**Department of Mathematics and Computer Science, University of Leicester and **Department of Computer Science, Technical University of Denmark*

Abstract Dependability analysis of imperfect (i.e. failure-prone) real-time systems or performance analysis of real-time systems interacting with stochastic environments is frequently based on the use of *Probabilistic Automata*, PA's, with Markov properties as computation models. A recent formalism using this approach is *Probabilistic Duration Calculus*, PDC, an extension of *Duration Calculus*, DC, which in turn was developed for specification and verification of embedded real-time systems. PDC makes it possible to calculate and reason about the probability that the PA satisfies a DC-formula (expressing a requirement or design decision) during the first t time units.

In this paper we consider parallel composition of component PA's into larger component PA's or into a system PA, where each component PA may depend on states in its environment (the PA is then said to be open), but where at least the system PA is independent of external states (the PA is then said to be closed).

We also consider probabilistic refinement with respect to a DC formula. This notion of refinement means that the probability that the present design (represented by a closed PA) satisfies the DC formula is greater than (or equal to) the corresponding probability for the previous design and the same formula.

Keywords: Probabilistic Automata, Markov Chains, Probabilistic Duration Calculus, Composition, Refinement, Dependability assessment.

---

[1] This research was supported by the Danish Technical Research Council under project Co-design. The research of Zhiming Liu was also supported in part by research grant GR/H39499 from the Science and Engineering Research Council of UK, when he was a research Fellow in the Department of Computer Science, University of Warwick

1 Introduction

In the design of embedded real-time systems common mathematical frameworks linking specification and verification with evaluation of probabilistic dependability or performance measures have received increasing attention recently [1, 2, 3, 4]. These approaches are all based on the use of probabilistic automata with Markov properties as computation models for the probabilistic evaluation.

In [1] the Markov model is constructed in such a way that it is consistent with local trace-based specifications. This model is then used for conventional numerical analysis of e.g. reliability or life-time requirements.

In contrast to this two-model approach [2, 3, 4] extend the formalisms used for specification and verification with probabilities, making it possible not only to calculate but also to reason about the probabilities that various requirements are satisfied.

[2] extends Computation Tree Logic [5] with probabilities. This approach is oriented towards investigation of soft deadline properties. Compared to this, improved expressiveness and reasoning power was obtained in [3, 4] which both extend Duration Calculus (DC) [6, 7, 8, 9] with probabilities.

The extended logic in [4], in the following called Probabilistic Duration Calculus (PDC), represents a simplification (preserving expressiveness) of the extended logic in [3]. PDC consists of a notation and a set of rules and theorems which make it possible to calculate and reason about the probability that the PA satisfies a DC-formula (expressing a requirement or design decision) during the first t time units.

Whereas compositionality and refinement are key issues in nonprobabilistic design techniques for real-time systems, these notions have only received minor attention in the probabilistic extentions of these techniques.

In this paper we extend the work in [4] by considering parallel composition of Probabilistic Automata (henceforth called PA's) into larger component PA's or into a system PA. The probability parameters defining a component PA may or may not depend on the environment of the PA (i.e. on states in other component PA's). If such dependencies exist, the PA is said to be open, otherwise it is said to be closed. Parallel composition of two component PA's is defined in such a way that it eliminates (hides) mutual dependencies. The final system PA is always closed. Closedness of a PA is a prerequisite for reasoning about its probabilistic properties by means of PDC.

We also consider probabilistic refinement with respect to a DC formula (expressing a requirement). This notion of refinement means that the probability that the present design (represented by a closed PA) satisfies the DC formula is greater than (or equal to) the probability that the previous design (represented by another closed PA) satisfies the same DC formula.

An extended version of this paper with proofs of theorems is available as a technical report [10]. A more intuistic and less mathematical description of the work is presented in [11].

2 Probabilistic automata as models of imperfect systems

In the modelling of imperfect real-time systems by probabilistic automata the question naturally arises: How do we define probabilistic automata representing components of a design and how do we compose these automata into a probabilistic automaton representing the entire system as a Markov chain ? In this section we provide answers to these questions.

2.1 States

Let A denote the set of primitive states for a proposed design of a system G. Assume that A is partitioned into n non-empty sets:

$$A = A_1 \cup \ldots \cup A_n, \quad A_i \cap A_j = \emptyset$$

such that each A_i is *local* to a component (agent) G_i in the sense that states in A_i can be changed only by G_i.

As a running example we consider a simple Gas Burner [4] consisting of a Burner (an abstraction of gas-valve, igniter and control box) and a Detector (an abstract device for detection of unburnt gas). The gas is turned on at time $t = 0$ and remains on. The flame may disappear spontaneously at any time. This results in leakage of unburnt gas until re-ignition is accomplished.

Example 1. *For a Gas Burner, composed from a Burner and a Detector, the Burner makes transitions between state Leak (meaning gas and no flame) and state ¬Leak (meaning gas and flame), whereas the Detector makes transitions between state Act and state ¬Act, where Act (active) means that the detector is able to detect leaking gas.*

For this system the set of primitive states and its partition is:

$$A = \{Leak, Act\}, \quad A_1 = \{Leak\}, \quad A_2 = \{Act\}$$

Let 2^A denote the *total state base* over A, i.e. the set of basic conjunctions[2] of primitive states in A. We define $2^\emptyset \triangleq \{1\}$. A nonempty subset $\mathcal{S}$

---

[2] A conjunction of primitive states in A is *basic*, if every primitive state in A or its negation, but not both, appears in the conjunction. A basic conjunction of A is sometimes called a *minterm* of A.

of 2^A is called a *state base* over A. We will use x, y, v, x', y', v', etc. to range over a state base. For state bases $\mathcal{S}_1$ and $\mathcal{S}_2$ over disjoint sets A_1 and A_2 respectively, we define the product

$$\mathcal{S}_1 \times \mathcal{S}_2 \triangleq \{x \wedge y \mid x \in \mathcal{S}_1, y \in \mathcal{S}_2\}$$

which is a subset of $2^{A_1 \cup A_2}$. For a subset B of A and a state $v \in 2^A$, let $v|_B$ denote the conjunction in 2^B obtained from v by removing any primitive state (or its negation) when this primitive state is not in B.

Example 2. *In the Gas Burner example,*

$$
\begin{aligned}
2^{A_1} &= 2^{\{Leak\}} &&= \{Leak, \neg Leak\} \\
2^{A_2} &= 2^{\{Act\}} &&= \{Act, \neg Act\} \\
2^A &= 2^{\{Leak, Act\}} &&= 2^{\{Leak\}} \times 2^{\{Act\}} \\
& && = \{Leak \wedge Act, Leak \wedge \neg Act, \\
& && \quad\; \neg Leak \wedge Act, \neg Leak \wedge \neg Act\}
\end{aligned}
$$

$$Leak \wedge \neg Act|_{A_1} = Leak$$

A component G_i can only change its own local states, i.e. states in 2^{A_i}. However, when the component interacts with other components, the possibility of a transition from local state v_{i1} to local state v_{i2} may depend not only on v_{i1} but also on external states, i.e. states in $2^{A \setminus A_i}$.

Example 3. *In our Gas Burner example we assume the following behaviour of the Burner*

- *The Burner starts from state $\neg Leak$ with probability $p_1 = 1$.*

- *Given that it is in state $\neg Leak$, it remains in that state for one time unit with probability p_{11} or it goes to state $Leak$ within one time unit with probability p_{12} where $p_{11} + p_{12} = 1$. These probabilities are independent of the state of the Detector.*

- *Given that it is in state $Leak$, it remains in that state for one time unit with probability p_{22} or it goes to state $\neg Leak$ within one time unit with probability p_{21} where $p_{22} + p_{21} = 1$. If the detector is in state Act, $p_{21} > 0$. Otherwise $p_{21} = 0$ (implying that $p_{22} = 1$).*

This reflects that p_{11} and p_{12} characterise the stability of the flame and, thus, are entirely independent of the Detector state, whereas p_{22} and p_{21} characterise the ability of the Burner to eliminate a Leak state, and this ability depends on the Detector state.

It follows from this discussion that we should model the probabilistic behaviours of a component under various environments. This idea is captured and elaborated in the following.

2.2 Open and closed probabilistic automata

Definition 4. *Let $\mathcal{A}$ be a finite set of primitive states, $\mathcal{S}$ be a state base over $\mathcal{A}$ (i.e. a nonempty subset of $2^{\mathcal{A}}$) and C be a finite set of primitive states such that $\mathcal{A} \cap C = \emptyset$. A C-open automaton over $\mathcal{A}$ is a 4-tuple $G(\mathcal{S}|C) = (\mathcal{S}, C, \tau_0, \tau)$ where*

- *τ_0 is the initial probability function $\tau_0 : \quad \mathcal{S} \longmapsto [0,1]$ such that*
$$\sum_{x \in \mathcal{S}} \tau_0(x) = 1.$$

- *τ is the (single step) transition probability function $\tau : 2^C \longmapsto [\mathcal{S} \times \mathcal{S} \longmapsto [0,1]]$ such that for any $v \in 2^C$ and $x \in \mathcal{S}$ $\sum_{x' \in \mathcal{S}} \tau(v)(x, x') = 1$.*

We call $\tau_0(x)$ the initial probability of x and $\tau(v)(x, x')$ the transition probability from x to x' under condition v.

Example 5. *It follows from Example 3 and Definition 4 that the Burner can be modelled by the C-open probabilistic automaton*

$$Burner(\{Leak, \neg Leak\} | \{Act\}) = (\{Leak, \neg Leak\}, \{Act\}, \tau_0, \tau) \text{ where}$$

$$\tau_0(\neg Leak) = p_1 = 1 \qquad\qquad \tau_0(Leak) = p_2 = 0$$

$$\tau(Act)(\neg Leak, \neg Leak) = p_{11} \qquad \tau(Act)(\neg Leak, Leak) = p_{12} = 1 - p_{11}$$
$$\tau(Act)(Leak, \neg Leak) = p_{21} \qquad \tau(Act)(Leak, Leak) = p_{22} = 1 - p_{21}$$

$$\tau(\neg Act)(\neg Leak, \neg Leak) = p_{11} \qquad \tau(\neg Act)(\neg Leak, Leak) = p_{12}$$
$$\tau(\neg Act)(Leak, \neg Leak) = 0 \qquad\quad \tau(\neg Act)(Leak, Leak) = 1$$

The open probabilistic automaton, *Burner*, is illustrated by its transition graph in Figure 1a.

Definition 6. *A C-open probabilistic automaton $G(\mathcal{S}|C) = (\mathcal{S}, C, \tau_0, \tau)$ is said to be **closed**, if for any $v, v' \in 2^C$ and for any $x, x' \in \mathcal{S}$, $\tau(v)(x, x') = \tau(v')(x, x')$.*

Thus, closedness reflects that the automaton is independent of its environments. It follows from the definition (and the fact that $2^{\emptyset} = \{1\}$, i.e. a singleton set) that any automaton $G(\mathcal{S}|C)$ is closed if $C = \emptyset$.

If $G = (\mathcal{S}, \emptyset, \tau_0, \tau)$ is a (closed) probabilistic automaton over the set of primitive states $\mathcal{A}$ and $G' = (\mathcal{S}, C, \tau_0, \tau')$, where $C \cap \mathcal{A} = \emptyset$ and $\tau'(v)(x, x') \stackrel{\Delta}{=} \tau(1)(x, x')$ for any $x, x' \in \mathcal{S}$ and any $v \in 2^C$, then G' is a closed automaton equivalent to G in the sense that each sub- automaton of G' corresponding to a condition $v \in 2^C$ becomes a copy of G.

Thus if $G'(\mathcal{S}|C) = (\mathcal{S}, C, \tau_0, \tau')$ is a closed automaton, then we can write it simply as the triple $G = (\mathcal{S}, \tau_0, \tau)$ where $\tau(x, x') = \tau'(v)(x, x')$ for any $x, x' \in \mathcal{S}$ and some $v \in 2^C$.

Example 7. *In our Gas Burner example we assume the following behaviour of the Detector*

- *The Detector starts from state Act with probability q_1 or from state $\neg Act$ with probability q_2 $(= 1 - q_1)$.*

- *Given that it is in state Act, it remains in that state for one time unit with probability q_{11} or it goes to state $\neg Act$ within one time unit with probability q_{12} where $q_{11} + q_{12} = 1$. These probabilities are independent of the state of the Burner.*

- *Given that it is in state $\neg Act$, it remains in that state for one time unit with probability q_{22} or it goes to state Act within one time unit with probability q_{21} where $q_{22} + q_{21} = 1$. These probabilities are independent of the state of the Burner.*

This reflects that the probabilities of failure or repair of the Detector are entirely independent of flame failures (causing leaks) in the Burner.

Therefore, the detector is the closed probabilistic automaton

$$Detector(\{Act, \neg Act\}|\{\}) = (\{Act, \neg Act\}, \tau_0, \tau) \ \ where$$

$$\tau_0(Act) = q_1 \qquad\qquad \tau_0(\neg Act) = q_2 = 1 - q_1$$

$$\tau(Act, Act) = q_{11} \qquad \tau(Act, \neg Act) = q_{12} = 1 - q_{11}$$
$$\tau(\neg Act, Act) = q_{21} \qquad \tau(\neg Act, \neg Act) = q_{22} = 1 - q_{21}$$

The closed automaton, *Detector*, is illustrated by its transition graph in Figure 1b.

2.3   Parallel composition of probabilistic automata

In this section we consider parallel composition of open probabilistic automata. In view of Definition 6 this theory also applies to cases where some or all of the involved automata are closed.

First we define a restricted form of parallel composition (Definition 8) and then we make a generalization (Definition 9).

Definition 8. *For $i = 1, 2$ let $G_i(\mathcal{S}_i|C_i) = (\mathcal{S}_i, C_i, \tau_0^{(i)}, \tau^{(i)})$, be a C_i-open probabilistic automaton over A_i and let $A_1 \cap A_2 = \emptyset$, $A_1 \subseteq C_2$ and $A_2 \subseteq C_1$. Then the **parallel composition** of $G_1(\mathcal{S}_1|C_1)$ and $G_2(\mathcal{S}_2|C_2)$ is the 4-tuple*

$$G_1(\mathcal{S}_1|C_1) \parallel G_2(\mathcal{S}_2|C_2) \triangleq (\mathcal{S}_1 \times \mathcal{S}_2, ((C_1 \setminus A_2) \cup (C_2 \setminus A_1)), \tau_0, \tau)$$

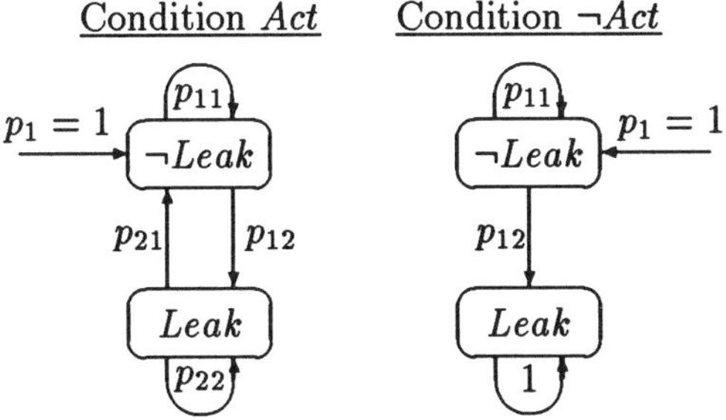

a: *Burner* (Open automaton)

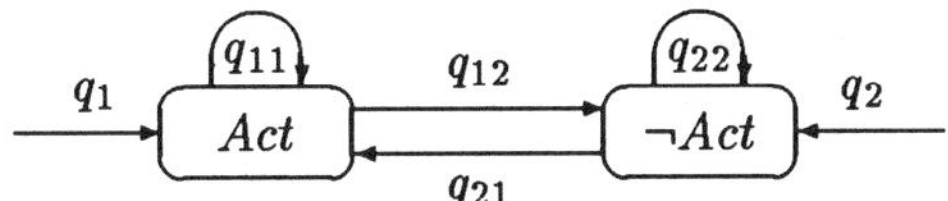

b: *Detector* (Closed automaton)

Figure 1. The probabilistic component automata: *Burner* and *Detector*

where for any $x_i, x_i' \in S_i$, $i = 1, 2$ and for any $v \in 2^{(C_1 \backslash A_2) \cup (C_2 \backslash A_1)}$

$$\tau_0(x_1 \wedge x_2) \triangleq \tau_0^{(1)}(x_1) * \tau_0^{(2)}(x_2) \ \ and$$

$$\tau(v)(x_1 \wedge x_2, x_1' \wedge x_2') \ \triangleq \ \tau^{(1)}(v|_{(C_1 \backslash A_2)} \wedge x_2)(x_1, x_1') * \\ \tau^{(2)}(v|_{(C_2 \backslash A_1)} \wedge x_1)(x_2, x_2')$$

For $i = 1, 2$ let $G_i(\mathcal{S}_i | C_i) = (\mathcal{S}_i, C_i, \tau_0^{(i)}, \tau^{(i)})$ be a C_i-open probabilistic automaton over A_i. We can then introduce $\overline{G}_i(\mathcal{S}_i | C_i \cup A_j) = (\mathcal{S}_i, C_i \cup A_j, \tau_0^{(i)}, \overline{\tau}^{(i)})$, $i, j = 1, 2$ and $i \neq j$, such that for any $x, x' \in S_i$ and $v \in 2^{C_i \cup A_j}$ $\overline{\tau}(v)(x, x') \triangleq \tau(v|_{C_i})(x, x')$. Notice that if $A_j \subseteq C_i$ then $\overline{G}_i(\mathcal{S}_i | C_i \cup A_j) = G_i(\mathcal{S}_i | C_i)$.

Definition 9. *For $i = 1, 2$ let $G_i(\mathcal{S}_i | C_i) = (\mathcal{S}_i, C_i, \tau_0^{(i)}, \tau^{(i)})$ be a C_i-open probabilistic automaton over A_i and let $A_1 \cap A_2 = \emptyset$: then the* **parallel composition** *of $G_1(\mathcal{S}_1 | C_1)$ and $G_2(\mathcal{S}_2 | C_2)$ is defined by $G_1(\mathcal{S}_i | C_1) \parallel G_2(\mathcal{S}_2 | C_2) \triangleq \overline{G}_1(\mathcal{S}_1 | C_1 \cup A_2) \parallel \overline{G}_2(\mathcal{S}_2 | C_2 \cup A_1)$*

The following theorem is proved in [10]

156　　　　　　　　　　　　　　　　*Z. Liu et al.*

Theorem 10. *Under the conditions of Definition 9 the parallel composition $G_1(\mathcal{S}_i|C_1) \parallel G_2(\mathcal{S}_2|C_2)$ is a $((C_1 \setminus A_2) \cup (C_2 \setminus A_1))$-open probabilistic automaton with state base $\mathcal{S}_1 \times \mathcal{S}_2$ over $A_1 \cup A_2$.*

Corollary. *If $C_1 \subseteq A_2$ and $C_2 \subseteq A_1$, then $G_1(\mathcal{S}_1|C_1) \parallel G_2(\mathcal{S}_2|C_2)$ is a closed probabilistic automaton.*

Example 11. *The Gas Burner continued.*
In Definition 9 let

$$
\begin{aligned}
G_1(\mathcal{S}_1|C_1) &= Burner(\{Leak, \neg Leak\}|\{Act\}) \quad &\textit{(cf. example 3)}\\
G_2(\mathcal{S}_2|C_2) &= Detector(\{Act, \neg Act\}|\{\}) \quad &\textit{(cf. example 5)}
\end{aligned}
$$

then, since $A_2 = C_1 = \{Act\}$ (implying that $A_2 \subseteq C_1$), $\overline{G}_1(\mathcal{S}_1|C_1 \cup A_2)$ becomes identical to $G_1(\mathcal{S}_1|C_1)$. Furthermore, in order to obtain $\overline{G}_2(\mathcal{S}_2|C_2 \cup A_1)$ we just have to replace $C_2 (= \{\})$ by $C_2 \cup A_1 (= \{Leak\})$ in $G_2(\mathcal{S}_2|C_2)^3$.

Having determined $\overline{G}_1$ and $\overline{G}_2$ we can calculate their parallel composition (defining the parallel composition of the original G_1 and G_2). The details of this calculation are specified by Definition 8. According to the corollary of Theorem 10 the resulting probabilistic automaton, henceforth called GasBurner, is closed.

The transition graph of GasBurner is shown in Figure 2. The reasoning behind the construction is illustrated informally by a few examples.

- *As shown in Figure 1, Burner starts in state $\neg Leak$ with probability $p_1 = 1$ while Detector starts in state Act with probability q_1 or in state $\neg Act$ with probability $q_2 = 1 - q_1$. This implies, that GasBurner $=$ Burner $\parallel$ Detector starts in state $(\neg Leak \wedge Act)$ with probability $p_1q_1 = q_1$ or in state $(\neg Leak \wedge \neg Act)$ with probability $p_1q_2 = q_2$.*

- *If Burner is in state $\neg Leak$ it can jump to state Leak within one time unit with probability p_{12}. If Detector is in state Act it can jump to state $\neg Act$ within one time unit with probability q_{12}. This implies that if GasBurner is in state $(\neg Leak \wedge Act)$ it can jump to state $(Leak \wedge \neg Act)$ within one time unit with probability $p_{12}q_{12}$.*

- *If Burner is in state Leak and Detector is in state $\neg Act$, Burner will remain in state Leak during the next time unit with probability 1 (because the leak can not be detected). The detector may stay in state $\neg Act$ within the next time unit with probability q_{22}. This implies that if GasBurner is in state $(Leak \wedge \neg Act)$ it may stay in that state within the next time unit with probability $1 \times q_{22} = q_{22}$.*

---

³This implies duplication of $\tau^{(2)}$ in $G_2(\mathcal{S}_2|\{\}) = (\mathcal{S}_2, \{\}, \tau_0^{(2)}, \tau^{(2)})$ for each condition $v \in 2^{A_1}$

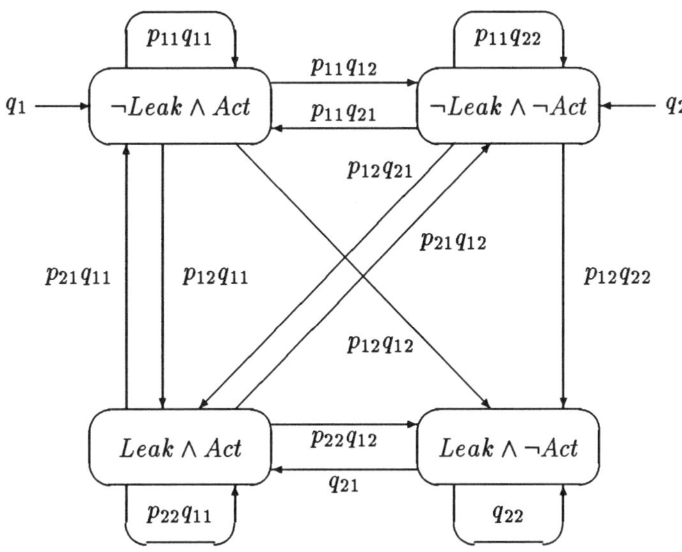

Figure 2. The closed automaton $GasBurner = Burner \parallel Detector$

The other transitions can be reasoned about in a similar way.

The following two theorems complete the theory of parallel composition of probabilistic automata. They are proved in [10].

Theorem 12. *Let G_1, G_2 and G_3 be open probabilistic automata. Then*

$$G_1 \parallel G_2 = G_2 \parallel G_1 \ \ and \ (G_1 \parallel G_2) \parallel G_3 = G_1 \parallel (G_2 \parallel G_3)$$

Theorem 13. *If G_1 and G_2 are closed probabilistic automata, then so is* $G_1 \parallel G_2$.

3 Specifying prob. automata with duration calculus

The duration calculus [6] was developed for specification and verification of real-time systems in terms of their states. In this section it is shown how to use a discrete time version of duration calculus as a specification language for probabilistic automata, and how to calculate the probability that a probabilistic automaton satisfies a formula in the duration calculus. The computation model is a simple Markov chain over discrete time (the corresponding model over continous time would be a semi - Markov process, which is much more difficult to analyse).

158   *Z. Liu et al.*

3.1   Duration calculus with discrete time

The duration calculus assumes a finite set A of primitive states. The primitive states are interpreted by a function $I : N_0 \longmapsto (A \longmapsto \{0,1\})$ where
time is modelled by the non-negative integers N_0, and $I(P)(t) = 1$ means
that the primitive state P is present at time point t. State expressions
can be formed from primitive states and the Boolean operators $\neg, \wedge, \vee, \Rightarrow$.
The interpretation function is extended to state expressions in the natural
way, e.g. $I(P \wedge \neg Q)(t) = 1$ if and only if $I(P)(t) = 1$ and $I(Q)(t) = 0$
For any state expression P, its duration $\int P$ (called a duration expression)
is interpreted over a time interval as the cumulated time during which
$I(P) = 1$:

$$I(\textstyle\int P)([t_1,t_2]) \triangleq \sum_{t=t_1}^{t_2-1} I(P)(t)$$

For a point interval $[t,t]$ the duration is defined to be zero: $I(\int P)([t,t]) \triangleq$
0.

A duration formula is an expression of Boolean type formed from duration expressions, integer variables, integer operators (e.g. $+,-,*$), relational operators on integers (e.g. $=,>,<$), Boolean operators, the chop
operator (;) to be defined below, and quantifiers $\forall, \exists$ applied to integer
variables. A duration formula D is satisfied by an interpretation I with an
interval $[t_1,t_2]$ just when it is evaluated to true for that interpretation and
that interval. This is written as $I, [t_1,t_2] \models D$.

Chop (;) is a dyadic infix operator taking two duration formulae and
yielding a new duration formula. An interpretation I with an interval $[t_1,t_2]$
satisfies $D_1; D_2$ if the interval can be divided in two coherent intervals, such
that D_1 is satisfied for the first interval and D_2 is satisfied for the second
interval:

$$I, [t_1,t_2] \models D_1; D_2 \quad \text{iff.} \quad \exists t' \in [t_1,t_2] \bullet (I, [t_1,t'] \models D_1 \wedge I, [t',t_2] \models D_2)$$

For duration formulae, the following notation is used: The symbol ℓ denotes
the length of the interval over which a formula is evaluated: $\ell \triangleq \int 1$. For
any state expression P, $\lceil P \rceil$ denotes that P holds everywhere in a non-point
interval.

$$\lceil P \rceil \triangleq (\textstyle\int P = \ell) \wedge \ell > 0$$

For any duration formula D, the formula $\Diamond D$ is satisfied for an interval if
D is satisfied for a subinterval.

$$\Diamond D \triangleq true; D; true$$

The formula $\Box D$ is satisfied for an interval if D is satisfied for all its subintervals.

$$\Box D \;\triangleq\; \neg\Diamond\neg D$$

3.2 Probabilistic satisfaction of duration calculus formulae

Let $G = (\mathcal{S}, \tau_0, \tau)$ be a closed probabilistic automaton over A, and let $\mathcal{S}^t, t \geq 0$ denote the set of sequences of length t of elements of $\mathcal{S}$. Each element of $\mathcal{S}^t$ is the state sequence corresponding to the first t time units of an execution of G. The probability of G executing $\sigma \in \mathcal{S}^t$ during the first t time units (given that $\sigma = \langle \sigma[0], \sigma[1], \ldots, \sigma[t-1]\rangle$) is defined as

$$\mu(\sigma) \;\triangleq\; \tau_0(\sigma[0]) * \prod_{i=1}^{t-1} \tau(\sigma[i-1], \sigma[i])$$

The probability of the empty sequence is defined as $\mu(\langle\rangle) = 1$. With these definitions it follows that $\langle \mathcal{S}^t, \mu\rangle$ is a probability space with $\mathcal{S}^t$ as sample space.

Now, let D be a duration formula. Each $\sigma \in \mathcal{S}^t$ defines an interpretation I_σ for which D can be evaluated. For a time point $t' \in [0, t-1]$ and primitive state x

$$I_\sigma(x)(t') \;\triangleq\; \textbf{if } \sigma[t'] \Rightarrow x \textbf{ then } 1 \textbf{ else } 0$$

For $t' \notin [0, t-1]$ the value of $I_\sigma(x)(t')$ does not matter. Let $\phi(D)$ be the set of sequences in $\mathcal{S}^t$ whose interpretations over the interval $[0, t]$ satisfy D:

$$\phi(D) \;\triangleq\; \{\sigma \in \mathcal{S}^t \mid I_\sigma, [0, t] \models D\}$$

Now we can define the probability that G satisfies D during the first t time units as

$$\mu(D)[t] \;\triangleq\; \sum_{\sigma \in \phi(D)} \mu(\sigma)$$

The fact that G satisfies D with probability $\nu(t)$ is denoted as

$$\mu_G(D)[t] = \nu(t)$$

A set of rules for reasoning about the probabilities of satisfying duration calculus formulae is given in [4].

Example 14. *In the GasBurner from the previous examples, the states Leak $\wedge$ Act and Leak $\wedge \neg$Act are critical because they correspond to release of unburnt gas. A requirement for the GasBurner could therefore be that whenever Leak is present, it should be detected and eliminated within one time unit. In duration calculus this is expressed as*

$$\textbf{Des-1}: \quad \Box(\lceil Leak \rceil \Longrightarrow \ell \leq 1)$$

The goal of probabilistic verification could be to establish that the probability of satisfying this requirement is sufficiently high:

$$\mu_{GasBurner}(\textbf{Des-1})[t] \geq Limit(t)$$

where $Limit(t)$ is the lower bound for acceptable probabilities.

3.3 Calculation technique

From the definition of $\mu(D)[t]$ it seems necessary to calculate $\phi(D)$ in order to calculate μ. However, for certain classes of duration formulas it is possible to calculate μ by manipulating the vector $\textbf{p}$ corresponding to the initial probability function and the matrix $\textbf{P}$ corresponding to the transition probability function [4]. For duration formulas D which can only be violated by initial states and single transitions, we have

$$\mu(D)[t+1] = \textbf{p}' \cdot (\textbf{P}')^t \cdot \textbf{1}_c$$

where $\textbf{1}_c$ denotes a column vector in which all elements are 1. The vector $\textbf{p}'$ is obtained from $\textbf{p}$ by zeroing all entries corresponding to states which will initially violate D. The matrix $\textbf{P}'$ is obtained from $\textbf{P}$ by zeroing all entries corresponding to transitions which will violate D.

Example 15. *The duration formula* **Des-1** *belongs to the class referred to above. Therefore, the probability $\mu_{GasBurner}(\textbf{Des-1})[t+1]$ can be calculated as follows: For the GasBurner we have*

$$\textbf{p} = (q_1,\, q_2,\, 0,\, 0) \ \ and \ \ \textbf{P} = \begin{pmatrix} p_{11}q_{11} & p_{11}q_{12} & p_{12}q_{11} & p_{12}q_{12} \\ p_{11}q_{21} & p_{11}q_{22} & p_{12}q_{21} & p_{12}q_{22} \\ p_{21}q_{11} & p_{21}q_{12} & p_{22}q_{11} & p_{22}q_{12} \\ 0 & 0 & q_{21} & q_{22} \end{pmatrix}$$

*where the numbering of rows and columns is arbitrarily chosen as follows:
1) $\neg Leak \wedge Act$, 2) $\neg Leak \wedge \neg Act$, 3) $Leak \wedge Act$ and 4) $Leak \wedge \neg Act$.*

Now, **Des-1** *cannot be violated by any initial state, because it takes two time units to violate* **Des-1**. *Therefore* $\textbf{p}' = \textbf{p}$. *Any transition from a Leak state to a Leak state violates* **Des-1**. *No other transitions can violate* **Des-1**. *Therefore* $\textbf{P}'$ *is obtained by zeroing entries (3,3), (3,4), (4,4) and (4,3). Accordingly,*

$$\mu_G(\textbf{Des-1})[t+1] =$$

$$(q_1,\, q_2,\, 0,\, 0) \cdot \begin{pmatrix} p_{11}q_{11} & p_{11}q_{12} & p_{12}q_{11} & p_{12}q_{12} \\ p_{11}q_{21} & p_{11}q_{22} & p_{12}q_{21} & p_{12}q_{22} \\ p_{21}q_{11} & p_{21}q_{12} & 0 & 0 \\ 0 & 0 & 0 & 0 \end{pmatrix}^t \cdot \begin{pmatrix} 1 \\ 1 \\ 1 \\ 1 \end{pmatrix}$$

4 Refinement of probabilistic automata

In non-probabilistic top-down system development, a specification on a high abstraction level is transformed stepwise into a more and more concrete specification of components and system architecture. Often the highest abstraction levels contains a certain amount of nondeterminism, corresponding to design choices which are yet to be made. With each step in the design, this nondeterminism is gradually removed, yielding more and more concrete (considered 'better' and 'better') specifications. This is captured formally by the notion of refinement: A specification S_1 refines a specification S_0 if S_1 is 'better' than S_0 in terms of being less nondeterministic.

In probabilistic top-down system development the motivation is often to improve the designed systems reliability with respect to the reliability of its components, e.g. by introducing redundancy in a transformation step. Therefore, we define a notion of probabilistic refinement by interpreting 'better' as being more reliable.

In our context, reliability is taken to be the probability of satisfying a duration formula expressing desired behaviour.

Definition 16. *Let G_0 and G_1 be closed probabilistic automata, and let D be a duration formula. Then G_1 is said to* refine G_0 *with respect to D if and only if*

$$\forall t \geq 0 \bullet \mu_{G_1}(D)[t] \geq \mu_{G_0}(D)[t]$$

Example 17. *In order to improve the reliability of the Gas Burner, its design is changed such that it uses two detectors instead of one. With the new design, the burner component will know about absence of the flame if at least one of the detectors is active. Only in the case where both detectors are inactive will the absence of flame be unnoticed. The two detectors are equal to the Detector from the previous examples except for renaming of the state Act:*

$$Detector_1 = Detector[Act_1/Act]$$
$$Detector_2 = Detector[Act_2/Act]$$

The burner component have to be changed in order to acommodate two detectors: Under the conditions $Act_1 \wedge Act_2$, $Act_1 \wedge \neg Act_2$ and $\neg Act_1 \wedge Act_2$ (i.e. the conditions corresponding to at least one detector being active) NewBurner will behave as the original Burner under the condition Act. Under the condition $\neg Act_1 \wedge \neg Act_2$, NewBurner will behave as the original Burner under the condition $\neg Act$.

$$NewBurner(\{Leak, \neg Leak\} \mid \{Act_1, Act_2\}) =$$
$$(\{Leak, \neg Leak\}, \{Act_1, Act_2\}, \tau_{0NB}, \tau_{NB})$$

where

$$\tau_{0NB} = \tau_0$$
$$\tau_{NB}(Act_1 \wedge Act_2) = \tau_{NB}(Act_1 \wedge \neg Act_2) = \tau_{NB}(\neg Act_1 \wedge Act_2) = \tau(Act)$$
$$\tau_{NB}(\neg Act_1 \wedge \neg Act_2) = \tau(\neg Act)$$

and where τ_0 and τ refers to the initial probability function and the transition probability function, respectively, of the original Burner.

The new and improved gasburner is defined by

$$NewGasBurner = NewBurner \parallel Detector_1 \parallel Detector_2$$

Intuitively, the probability of satisfying **Des-1** *is higher for NewGasBurner than for GasBurner. Therefore we expect that NewGasBurner refines GasBurner with respect to* **Des-1***. This can be validated by calculating $\mu(\textbf{Des-1})[t]$ for NewGasBurner and GasBurner for a suitable range of t's.*

5 Conclusion

We have presented new results concerning composition, analysis and refinement of probabilistic real time systems. These notions are intended to support the development of such systems.

Our rules for parallel composition account for the cases where the stochastic behaviour of a component depends on internal states in other components.

Our notion of probabilistic refinement compares two designs with respect to their reliability by interpreting 'better' as meaning more reliable. In this context 'reliable' means the probability of satisfying a desirable property over a specified time interval.

The technique needs further consolidation with regard to tools and theorems for validation and verification, and its practical applicability to realistic dependability problems remains to be tested.

Bibliography

1. Sørensen, E.V., Nordahl, J. and Hansen, N.H. (1993). From CSP models to Markov models. *IEEE Transactions on Software Engineering, Vol. 19, No. 6*, 554-70.

2. Hansson, H. and Jonsson, B. (1989). A framework for reasoning about time and reliability. *Proceedings of the Tenth IEEE Symposium on "Real-Time Systems"*, Santa Monica, California, USA.

3. Liu, Z., Ravn, A.P., Sørensen, E.V. and Zhou, C. (1993). A probabilistic duration calculus. *Dependable Computing and Fault Tolerant Systems, Vol. 7*, Responsive Computer Systems, Springer Verlag, 29-52.

4. Liu, Z., Sørensen, E.V., Ravn, A.P. and Zhou, C. (1993). Towards a calculus of systems dependability. *Journal on High Integrity Systems*, Oxford University Press, (to appear).

5. Clarke, E.M., Emerson, E.A. and Sistla, A.P. (1983). Automatic verification of finite-state concurrent systems using temporal logic specifications: A practical approach. *Proceeding of the Tenth ACM Symposium on "Principles of Programming Languages"*, pp. 117-26.

6. Zhou, C.C., Hoare, C.A.R. and Ravn, A.P. (1991). A calculus of durations. *Information Processing Letters, Vol. 40, No. 5*, 269-76.

7. Ravn, A.P., Rischel, H. and Hansen, K.M. (1993). Specifying and verifying requirements of real-time systems. *IEEE Transactions on Software Engineering, Vol.19, No.1*, 41-55.

8. Ravn, A.P. and Rischel, H. (1991). Requirements capture for embedded real-time systems. *Proceeding of IMACS-IFAC Symposium MCTS*, Lille, France, pp. 147-52.

9. Skakkebæk, J.U., Ravn, A.P., Rischel, H. and Zhou, C.C. (1992). Specification of embedded real-time systems. *Proceedings of the Fourth Euromicro Workshop on "Real-Time Systems"*, IEEE Computer Society Press, pp. 116-21.

10. Liu, Z., Nordahl, J. and Sørensen, E.V. (1993), Composition and refinement of probabilistic automata. *Technical Report*, Department of Computer Science, Technical University of Denmark, 1-14.

11. Liu, Z., Nordahl, J. and Sørensen, E.V. (1993), Composition and refinement of probabilistic real-time systems. *Proceedings SAFECOMP'93*, Poznan-Kierkrz, Poland, Springer Verlag, pp. 30-40.

Fault prediction for software development processes[1]

J. May*, P. Hall*, H. Zhu*, T. Cockram, N. Bird[†] and L. Winsborrow[††]**

**Department of Computing, Open University and Nuclear Electric,*
***Rolls Royce, [†]Lucas Engineering and Systems and [††]Nuclear Electric*

Abstract Most activities during the software development process have the potential to cause faults in the final software code. A fault caused somewhere in the development process will be manifested in products produced after its introduction e.g. a high-level design document, or a code component. A probabilistic model of fault introduction is presented, with the aim of estimating the integrity of the work carried out up to a particular point in a given development process. In particular, estimates of the integrity of products early in the development lifecycle are sought.

1 Introduction

Faults in software code can be caused by errors at any stage in the software development lifecycle for that code. They are discovered by reviews, and later in the lifecycle by testing. The efficacy of a review depends on, amongst other factors, the review effort expended. However, there is currently no method of deciding how much effort should be expended on reviews. There is also no method for deciding where best to concentrate this effort within the development lifecycle.

Current thinking emphasizes the importance of ensuring the integrity of products in the early stages of the development lifecycle [1] [2]. Errors made early in the lifecycle can lead to a large amount of corrective work (debugging) later. Testing can only occur in the latter stages of the development lifecycle, and so reviews are the only means of uncovering faults in these critical early lifecycle stages. In theory faults are exposed by reviews, and are corrected prior to production of code. In practice reviews can not

---

[1]Research conducted within FASGEP (Fault Analysis of the Software Generation Process), a DTI/SERC SAFEIT project, no. IED4/1/9004, jointly by Nuclear Electric Plc., Lucas Engineering and Systems Limited, Lucas Electronics Limited, Lloyd's Register, Rolls Royce, and the Open University

be expected to find all faults. Furthermore, if reviews identify the need for extensive re-development of existing work, this is itself problematic, and sometimes may even introduce more errors than it removes [3].

Code debugging remains notoriously difficult and expensive [1]. Therefore, in view of the above, a method of monitoring the integrity of software development work *during* the development would be of great practical use. If such a method could recognize low integrity work soon after it occurs, reviews could be intensified to rectify matters. It would be particularly important that such a method could be applied during the early stages of the software development lifecycle. To this end, this paper studies the dependence of the integrity of products in the software development lifecycle (mainly software designs at various levels of abstraction) on immediately observable features of the development activities and the products themselves. Intuitively, a huge variety of such features may influence the number of faults introduced into such products, and how many of them remain after some review work has occurred e.g. the type of development process used, the task and product complexities, the time available, and the personnel involved. Such process/product features will be called *attributes*. The FASGEP project has developed probability models of the relationship between attributes and fault numbers. It is hoped to show that probabilistic estimation of fault numbers is possible, by verifying the model's estimates against real fault data. This paper reports the probability modelling techniques used. It was found that a novel modelling approach was required for this estimation problem.

Since there is always an unknown number of remaining faults in software, the FASGEP model cannot predict this number of faults. There would be no way to verify the model[2]. We actually observe faults which are found, whether during reviews, testing or software usage. It is the numbers of these 'findable' faults which the model predicts. It should be remembered that such a prediction occurs at an early stage in the development lifecycle i.e. before all of these faults have been observed. Of course a tool based on such predictions will fail to recommend review action when the total number of faults remaining is high, but the number of findable faults remaining is low. Nevertheless, the ability to predict the extent of the remaining findable faults is sufficient information for the purpose required, namely, to provide some guidance on the employment of review effort throughout the development lifecycle. This is because if the number of such faults is high then review effort should certainly be expended. In the above scheme, model verification appears problematic. This is because the total number of faults found will depend on the degree of reviewing, testing and usage which has been carried out, and this will vary between different

---

[2]Except possibly using fault seeding techniques, but this is beyond the scope of the paper

development projects. However, the data from Adams' experiment suggests that above a certain level of fault-finding effort, the number of faults found is altered little [15]. Thus the verification of the FASGEP model (Section 5) assumes the law of diminishing returns suggested by Adams' data, and also assumes that the model is verified using projects which employ enough fault-finding effort to make the assumed law applicable.

2 Software development processes

Existing models of software development processes usually consist of a set pattern of set activities. For example the standard 'V' model [4] is of this type. In contrast, the process models used here employ set activities organized in patterns which are not pre-set, and which depend on the software being developed. These patterns correspond closely to the structure of the particular software design as it evolves. That is, it was assumed that the development process can be decomposed in a manner corresponding to the software architecture, and that the resulting form of the process, its **skeleton**, *is a significant influence on the integrity of products produced by the process*. The approach to modelling this influence, in particular the treatment of corrective work, makes the FASGEP model highly novel. The following two sections describe the FASGEP approach to process modelling for the purposes of integrity estimation.

2.1 Process structure

The software development process is considered as repeated application of design, review, and integration activities, occurring in serial and parallel combinations which can be represented by a directed acyclic graph (DAG). Each activity has a set of input and output products associated with it. A design activity is the division of a specification into a 'design' i.e. a set of component specifications together with a specification of communications between the specified components. The object of a design activity is to obtain a design which is equivalent to the specification entering the design activity. A review activity is the inspection (or other means of error-checking) of a product, possibly resulting in corrective work for faults identified by the review. Review activities do not include the corrective work; faults are corrected by replacing all or parts of old design activities. Integration activities concern the fusion of products into one product, according to a specification of how they should be fused. For example, the integration of two code components; this activity would comprise the coding of the communications between the components (the components' *connectivity*). Similarly, the fusion of two design documents by straightforward concatenation would be an integration activity.

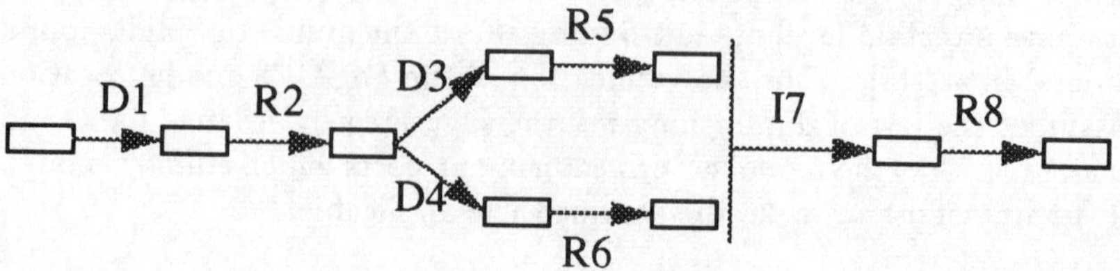

Figure 1. A simple process skeleton DAG

An extremely simple example of a software development process skeleton is represented in Figure 1. The overall process can be considered as combinations of various smaller processes i.e. sub-graphs. Processes corresponding to individual activities are represented by single edges in the DAG, which will be called *atomic processes*, or *atoms*. The atoms in Figure 1 are named D1, R2, D3, D4, R5, R6, I7, and R8. All other processes (any connected configurations of more than one atom) are *composite processes*.

The rectangles at the nodes in Figure 1 denote products. Products exist at every junction between atoms, but may be omitted in subsequent figures for convenience. Note that all atoms take one input and possess one output, except integration atoms (e.g. I7 in Figure 1) which may take any number of inputs and produce one output. The simple process skeleton in Figure 1 might be interpreted as follows. A design D1 specifies two components together with a specification of how they communicate. The results of D1 (the design documents containing the details of the design) are subjected to review in R2, which finds no faults. D3 and D4 code the components specified in D1 directly. Each code component is subjected to a review (R5, R6 respectively neither of which finds any faults). The two components are then integrated in atom I7, according to the communications specification from D1. Finally, a code walkthrough R8 finds no faults.

2.1.1 Modelling review effects

Reviews do not affect their input products directly. When depicted on the process skeleton the input and output products of reviews are always the same entity. Reviews are only included in a skeleton to show that a review was conducted at a certain point in the process. However, if a review identifies a fault, corrective work is initiated to remove the fault. For example, suppose R6 (Figure 1) identifies a fault in the design which was caused in D1. Then, if no other reviews found faults, a skeleton such as Figure 2 might result.

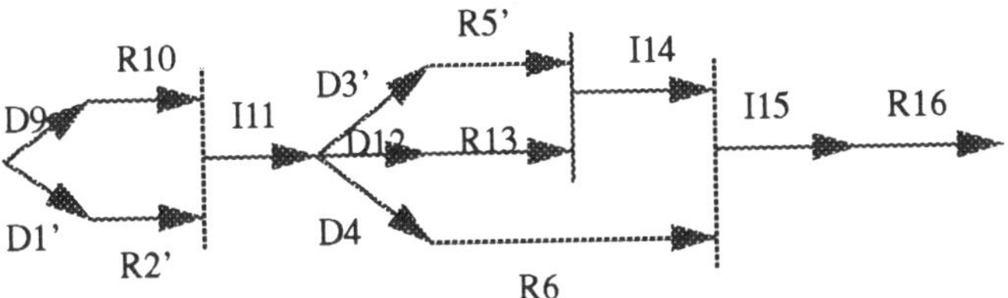

Figure 2. A process skeleton including a model of a particular fault correction

The DAG in figure 2 illustrates the fact that the process of corrective work can require a complex skeleton, which is consistent with the discussion in Section 1, corrective work is often a source of further faults. Complex skeletons are particularly likely when reviews discover faults caused by design work which does not directly precede the review (i.e. is earlier in the process) as in the example chosen for Figure 2. The form of the skeleton representing corrective work depends on the nature of the correction: the entities being corrected, and the processes used to achieve the corrections. The graph in Figure 2 represents an example of corrective work as described below. R6 identified a fault in the first component specification produced in D1. The specification of the component communications produced in D1 was unaffected. This leads to the following interpretation of Figure 2 in which A' denotes "part of the activity which was denoted by atom A", and is described as a *denuded* atom.

D9   is a new design process which re-specifies the faulty part of the specification for component 1 in the original design; and specifies how to connect the newly specified part to the old non-faulty part of the original component 1 specification i.e. D9 could be viewed as re-specifying the first component as two sub-components and their connectivity.

D1'   represents the design work from D1 which remains valid: the specification of component 2; the non-faulty part of the component 1 specification; and the specification of the two components' communications.

R10   is a review of the work produced in D9.

R2'   represents the review work from R2 which remains valid, namely, that done on the product of D1'.

I11   is a combination of the products from D1' and D9 to produce a single design document.

D3'   represents the parts of the coding of component 1 which correspond to the parts of its specification unchanged by the corrective work.

D12   represents the new coding replacing the parts of the old component 1 code which corresponded to the faulty parts of the component 1 specification i.e. D12 implements code according to the replacement specification from D9.

I14   implements the connection between the new code and the old code for component 1, as specified in D9.

I15   is the same process as I7 in Figure 1, since the specification of the connectivity of components 1 and 2 was unaffected by the fault.

The skeletons in Figure 1 and Figure 2 represent whole development processes, albeit simple ones. In practice, these skeletons would be built up dynamically as the development process proceeded for a particular piece of software. Thus in the above scenario, the process might have reached R5 and R6 and therefore be represented by the skeleton in Figure 1 but up to R5 and R6 only[3]. Then the identification of the above fault and all of the associated corrective work would be represented by the skeleton of Figure 2 up to atoms I14 and R6 only.

2.2   Immediately observable attributes

Each development and integration atom has process/product attributes associated with it (see Section 1). All development and integration atoms have the same set of attributes. As each new atom is added to the skeleton, a set of attribute measurements is made for the activity represented by the atom. The FASGEP model contains in excess of 100 attributes.

2.3   Faults, null processes, and process snapshots

A fault is uniquely associated with one atom: the one representing the activity in which the fault was caused. However, a fault can be said to exist in many products. If a fault is caused in atom A, then it exists within the product P at the output of A, and also in a subset of the products along paths leading away from P[4]. To see why a fault only exists within a subset of products on paths from the first product it appears in, consider an undiscovered fault caused in D1 of Figure 1. It will be present in the output product of D1 and R2 but, since its precise nature is not known, it cannot meaningfully be attributed to either of the output products for D3

---

[3]i.e. following the directed edges from the root or initial node

[4]A path exists from product P1 to product P2 if there is a directed edge from P1 to P2, or if there is a directed edge from P1 to P3 and a path from P3 to P2

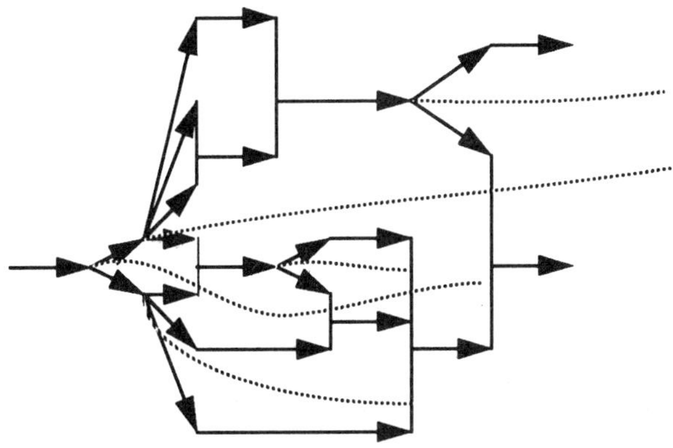

Figure 3. Examples of null processes

or D4. However, it will be present in the output products of I7 and R8. That is, a particular view of faults is taken. A fault caused early in the process is not considered to 'fan out' into many derivative faults. This is because it is not clear how such derivative faults could be defined; consider code derived from a specification fault, is each code instruction considered to constitute a fault?

To explain how faults are considered to accumulate in the FASGEP model, it is convenient to introduce the null process.

Definition 1. Null Process *A diverging node is one with more than one arrow pointing away from it. A null process is an atom whose output product is the same as its input product, and which represents no activity whatever. Null processes can only appear in processes at certain positions. A single null process emanates from each diverging node. A null process can only end with its output product at an input to an integration process. A given null process emanating from node N ends at the first integration process at which all paths emanating from N have converged [i.e. some of these paths may have converged earlier in the skeleton]. If there is no such integration process the null process is open-ended.*

Examples of null processes are shown as dotted lines in Figure 3, which has 5 diverging nodes, 2 of these emanate open-ended null processes.

Let F_X denote the number of faults present in product X, if X denotes a product. Similarly, let F_Y denote the number of faults introduced during process Y, if Y denotes a process. There are three cases of fault accumulation, as follows.

Serial processes

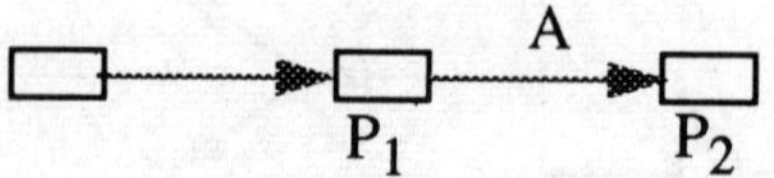

$$F_{P_2} = F_{P_1} + F_A$$

Diverging processes

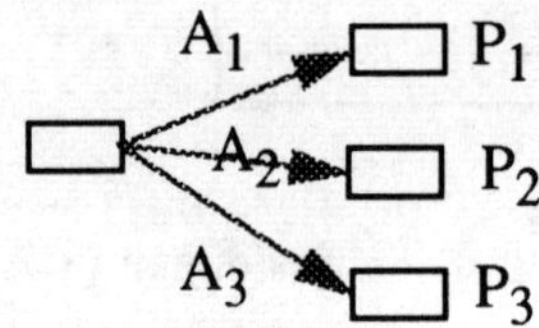

$$F_{P_i} = F_{A_i}$$

Converging processes

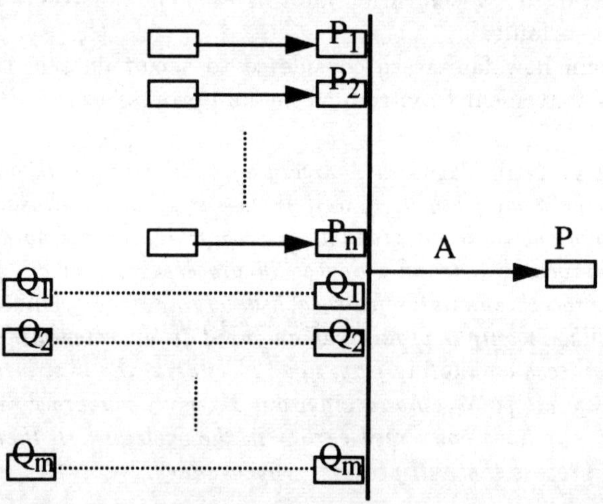

(The Qi are input and output products of null processes ending at the
input to integration process A.)

$$F_P = F_{P_1} + \cdots + F_{P_n} + F_{Q_1} + \cdots + F_{Q_m} + F_A$$

A main objective of FASGEP is the ability to estimate the number of faults in a set of products which represents "all work to date" in a development project. This cannot be achieved by simply summing the faults in the set of all products to date, since a fault can exist in more than one product. In Figure 2, immediately after completion of the activity for atom I11, it is sufficient to estimate the number of faults in the single product at the output of I11, since it contains all faults introduced in activities prior to I11.

To capture the notion of a product set which captures all work to date, the idea of a *process snapshot* is introduced.

Definition 2. Process snapshot for process P

A process snapshot S is a set of products in P such that:-

1. *all residual faults are represented in S i.e. exist within a product of S;*

2. *all faults in the products of S exist in one product of S only.*

Point (1) in Definition 2 implies that all paths from the root product pass through S. A snapshot could be thought of as a work 'frontier,' representing all the information needed to continue with the project. Thus a snapshot could, in principle, represent the state of the process at a point in time, although this is not a necessary condition for a set of products to be a snapshot.

3 A mathematics for monitoring integrity during development

3.0.1 Integrity progression

The term *reliability growth* is used to describe the trend of a program's integrity as the program is subjected to repeated testing/debugging cycles [5] [6]. The FASGEP techniques for integrity estimation aim to model a comparable phenomenon, but relating to fault numbers rather than reliability, and during much earlier stages in the development lifecycle. This phenomenon could be called *integrity progression*, since it is not likely to exhibit the growth of integrity over time which is expected during reliability growth estimation. Instead, a 'Christmas tree' effect is anticipated, as shown in Figure 4.

Integrity in Figure 4 is shown as a single magnitude which is the *expected value* of the fault propensity for a snapshot containing the set of products representing the current state of work. The Christmas tree effect occurs because the number of faults is likely to increase as design and integration

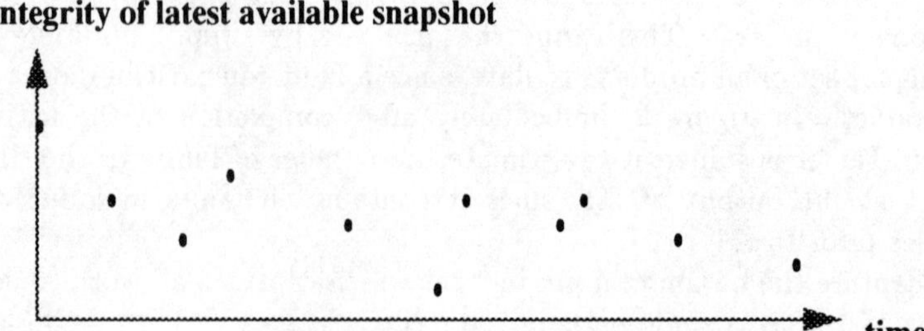

Figure 4. Integrity progression during software development

activities progress (causing a reduction in integrity), and is reduced by corrective work after reviews which have identified faults. However, it should be noted that the FASGEP model does not in any way assume that corrective work results in a reduction in overall fault numbers. Whether the model predicts an overall increase or decrease depends on the number of faults identified in the review and the fault propensities of the corrective processes.

3.0.2 An overview of the FASGEP model

The objective is a model to produce a probability distribution for the number of faults present in any given process snapshot. A probability distribution for the number of faults in any set of products will be termed the *fault-propensity* (fp) of that set of products[5].

The FASGEP model can be considered as two separate models which interact. The inner model (or Fault Introduction Model) models the relationship between observable process attributes (Section 1) for an atom and the number of faults introduced by that atom. An outer model (or Process Analysis Model) models the propagation of faults into products as the development process proceeds. That is, the outer model takes the fp for each atom in a process skeleton, as produced by the inner model, and combines these fps to produce an fp for each product in the process skeleton, and for any process snapshot of those products. The effects of reviews are modelled in the outer model.

---

[5]Fault-propensities are also associated with atoms, in which case they are a probability distribution for the number of faults introduced in the atom

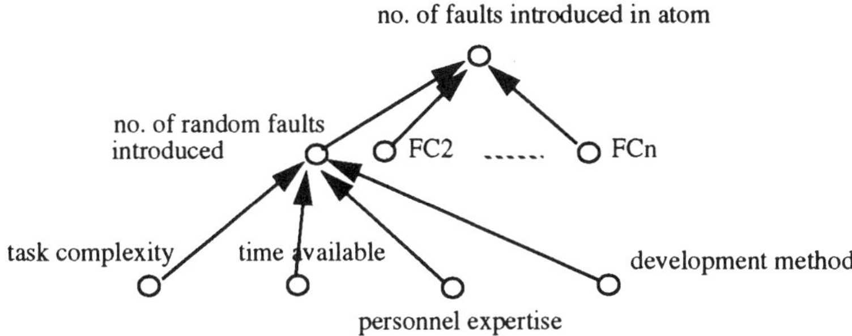

Figure 5. A causal network

3.1 The inner model

The FASGEP model requires that the fp for each individual design or integration atom be estimated. The FASGEP inner model performs this task, based on observations of the attributes (Section 1 and Section 2.2) of an atom. The relationship between attributes and fps is modelled using *graphical probability models* (GPMs). These probability models are described in detail elsewhere [7] [8] [9], and the subject is too large to cover here. For the purposes of this paper a brief description will suffice.

GPMs consist of a set of random variables and an associated set of probabilistic dependence relationships which hold amongst those variables. A GPM can be represented using a directed acyclic graph or DAG (known as a *causal network*), where the random variables of the GPM appear at the DAG nodes, and the structure of the DAG represents the dependence relationships. A simplified causal network is shown in Figure 5, in which FCi stands for 'number of category i faults introduced,' and which is only partially developed.

The immediately observable attributes appear as variables at the nodes of the graph, typically at the leaf nodes. Other nodes contain variables which are not observable at the time of the activity, and whose values at that time only become known at a later time. In particular, the top node in the FASGEP graph contains the variable whose value is to be predicted i.e. the probability distribution (in this case a fault propensity) of the variable is required. The same GPM is used for all atoms. Thus the fp for an atom, as predicted just after completion of the atom, depends only on that atom's immediately observed attribute measurements. In general causal networks for GPMs are DAGS, not simply tree structures as in Figure 5, and this is the case in the FASGEP model. The causal networks used in FASGEP contain in excess of 100 nodes, and are implemented using the HUGIN tool [10].

3.2 The outer model

The FASGEP outer model computes fps for process snapshots using fps of the atoms which produced those products. The fps for atoms are initially obtained from the inner model. The task of the outer model can be described in two parts. Firstly, alteration of fps for atoms which have been denuded as a result of corrective work. Secondly, propagation of fps through process skeletons when all denuding is completed. These two aspects of the outer model will be described in Section 3.2.1 and Section 3.2.2 respectively.

3.2.1 Fault propensity for denuded processes

For a description of denuded atoms see Section 2.1.1. The following assumption is made.

Assumption 1. Identified faults are removed *All faults identified by review no longer exist after the corrective work to remove them. These faults are either removed or transformed to other faults.*

The expected value of the fault-propensity of a denuded atom A' is reduced compared to that of the atom A of which it previously formed a part (Section 2.1.1). This is because a denuded atom A' represents a strict subset of the activity for atom A, and as such can only introduce less faults than A. If a review identifies m faults introduced during A, then the fp of A' is modelled by Equation 3.1, in which $fp_B(z)$ denotes the probability of atom B introducing z faults.

$$fp_{A'}(x) = \begin{cases} fp_A(x+m), & x > 0 \\ \sum_{y=0}^{m} fp_A(y), & x = 0 \end{cases} \tag{3.1}$$

3.2.2 Propagating fault propensity through process skeletons

All calculations are based around the convolution of fp functions, as given in Equation 3.2. The fp for a process snapshot containing products $P1, P2, \cdots, Pm$ is given by $\mathrm{conv}[fp_1, fp_2, \cdots, fp_m]$ in Equation 3.2.

$$\mathrm{conv}[fp_1, fp_2, \cdots, fp_m](n) = \sum_{n_1+n_2+\cdots+n_m=n} fp_1(n_1) fp_2(n_2) \cdots fp_m(n_m) \tag{3.2}$$

The computation of fps for individual products is based on Assumption 1, and Assumptions 2-4, which are Markov assumptions.

Assumption 2. Conditional independence of fault introduction
Conditional on the number of faults in each input product to atom A (including input products from null processes[6]), and the attributes for A, the number of faults introduced by A is independent of all other quantities in the model i.e. independent of any collection of attributes for atoms, independent of any fault numbers for products and/or atoms, and independent of any combination of attributes/fault numbers.

Assumption 3. Conditional independence of faults in products
Conditional on the number of faults in each input product to atom A, and the number of faults introduced by A, the number of faults in the output product P of atom A is independent of all other quantities in the model.

Assumption 4. Conditional independence of diverging processes
Conditional on the attributes of (non-null) atom A leading from a diverging node N, the number of faults in the product at the end of A is independent of all other quantities in the model (in particular is independent of the number of faults in the product at N).

The above Assumption 2, Assumption 3 and Assumption 4 imply the convolutions for the three process configurations in Section 2.3 as follows, in Section 3.2.2.1 to Section 3.2.2.3, Equation 3.3 to Equation 3.5 respectively. A derivation of Equation 3.3 to Equation 3.5 is given in [11]. Equation 3.2 to Equation 3.5 form the outer model together with the initial condition that the root node in the process skeleton is always assumed to have fp : $\{(0,1), (1,0), (2,0), (3,0), (4,0), \cdots\}$.

3.2.2.1 Serial processes

$$fp_{P_2} = \text{conv}[fp_{P_1}, fp_A] \tag{3.3}$$

3.2.2.2 Diverging processes

$$fp_{P_i} = fp_{A_i} \tag{3.4}$$

3.2.2.3 Converging processes

$$fp_p = \text{conv}[fp_{P_1}, \cdots, fp_{P_n}, fp_{Q_1}, \cdots, fp_{Q_m}, fp_A] \tag{3.5}$$

---

[6]Section 2.3

4 Experimental calibration of the model

The process of fitting the model to data is that of fitting a GPM to data, and is quite different in nature to classical statistical estimation. We call this fitting process 'calibration'. Calibrating the FASGEP model only concerns the inner model (i.e. the GPMs) since the outer model contains no parameters. The outer model is simply correct subject to the real software development process possessing the structure of the outer model process skeleton, and the validity of Assumption 1 to Assumption 4. It is the inner model which possesses parameters which can be modified to fit data. The inner model has two types of 'parameter':-

1. its graph structure;

2. conditional probability tables at the graph nodes.

GPM graph structure is usually considered fixed, and this is the case in the FASGEP model. The conditional probabilities are initially specified by experts according to a process which ensures their coherence [8]. However, these specified probabilities can then be treated as priors and updated as real data arrives in the form of coincident measurements of values taken by the variables of the GPM [12] [13]. The inner model must work under a very wide range of circumstances, and this means that calibration requires a large amount of data. Fortunately, large amounts of data will be available for the inner model: a reasonably sized software development project will comprise 100-1000 atoms. For each of these atoms, attribute observations, and an observed number of faults introduced during the atom, will be available. If initial results are encouraging, calibration of the FASGEP model (or different candidate models of the FASGEP type) will be an ongoing task as data is collected from additional projects. The FASGEP project is currently collecting data on four projects.

5 Experimental validation of the model

Data collection for the FASGEP project is already underway. To test the FASGEP model two types of data are required:-

1. observations of immediately observable attribute values for each atom in the process;

2. fault logs from all reviews, testing, and usage of the software, specifying the atoms in which the faults were introduced.

Once the FASGEP model has been calibrated, validation is a simple matter of comparing the model's fp distributions for each product in the outer model skeleton with the number of faults found for those products. The problem of assessing how well the inner model is predicting fault numbers has already been addressed in the literature: the use of score rules for measuring the performance of GPMs is reported in [14]. The performance of the overall model can be assessed using the same techniques.

6 Conclusions and further work

A mathematics enabling software development integrity to be monitored has been proposed. The modelling is radically different from other approaches to integrity prediction and, if successful, is the basis for a tool offering a new type of guidance during software development. The approach is currently being tested in the FASGEP project. The probability modelling is in two parts:-

1. *the inner model*, based on the existing framework of Graphical Probability Models (GPMs), and which models the introduction of faults during small sub-processes of the development process known as atoms;

2. *the outer model*, a bespoke probability model based on decomposition of the software development process into atoms, which combines the outputs of the inner model for all atoms and also models the effects of reviews and their associated corrective work, in order to estimate the fault propensity of software products.

The technique allows integrity estimates to be made as development proceeds and new products are produced. Thus "integrity progression" can be plotted in an analogous fashion to reliability growth. However, the two techniques have different uses. Reliability growth is studied where full working testable code is available, FASGEP models integrity progression in earlier stages of the software development lifecycle. Further, reliability growth is used to assess whether the software is of an acceptable integrity for release. The main use of integrity progression will be to influence the software development as it occurs, flagging unacceptable integrity conditions early, so that measures can be taken to rectify the situation.

If the FASGEP model can produce accurate predictions, future studies will include methods for using the model to assess the efficacy of candidate process forms in constructing software solutions. This will involve proposing processes in advance, and hence will require a technique for estimating the effects of reviews in advance. For this purpose a notion of review adequacy is being developed. Review adequacy will be estimated using a GPM, much as the current FASGEP GPMs estimate fault-propensities for atomic processes. The FASGEP model would also enable quantitative evaluation of the consequences (for integrity) of development practices of any kind which can be clearly defined in terms of the process attributes used in the model. Currently, only qualitative assessment of development practices is possible.

The FASGEP model is proposed as a first step towards evaluating the feasibility of monitoring a form of integrity (fault numbers) during software development. A novel modelling approach has been derived for this purpose. However, the immaturity of the approach means that it will probably need to evolve before predictions are sufficiently accurate. For example, more complex models of processes may be needed such as models of dependence between the integrities of processes conducted in parallel. The immediate aim of the FASGEP project is the assessment of the models which have already been produced i.e. an assessment of the accuracy of their predictions. The FASGEP project has a further year to run, and will be able to assess the promise of its approach in that time scale, using the very clear validation procedures described in this paper.

Bibliography

1. Sommerville, I. (1992). *Software Engineering (fourth edition)*.

2. Potter, B., Sinclair, J., and Till, D. (1991). *An Introduction to Formal Specification and Z.*, Prentice-Hall.

3. Myers, G. (1979). *The Art of Software Testing*, Wiley, New York.

4. McDermid, J. (1991). *Software Engineers Reference Book*, Butterworth-Heinemann.

5. Littlewood, B. (1988). Forecasting software reliability. *Software Reliability Modelling and Identification*, Editor: S. Bittanti, Springer-Verlag, Berlin.

6. Brocklehurst, S. and Littlewood, B. (1992). New ways to get accurate reliability measures. *IEEE Software*.

7. Pearl, J. (1988). *Probabilistic Reasoning in Intelligent Systems: Networks of Plausible Inference*, Morgan Kaufmann, San Mateo.

8. Lauritzen, S.L. and Spiegelhalter, D.J. (1988). Local computations with probabilities on graphical structures and their application to expert systems. *J. Royal Statistical Society B, 50, No. 2.*

9. Spiegelhalter, D.J., Dawid, A.P., Lauritzen, S.L. and Cowell, R.G. (1993). Bayesian analysis in expert systems. *Statistical Science, 8.*

10. Anderson, S.K., Jensen, F.V. and Olensen, K.G. (1987). The HUGIN core - preliminary considerations on systems for fast manipulations of probabilities. *Proceedings of Workshop on Inductive Reasoning: Managing Empirical Information in AI-systems.*

11. May, J.H.R. (1992). Analysis techniques. FASGEP Report No. NE06R01C.

12. Spiegelhalter, D.J. and Lauritzen, S.L. (1990). Sequential updating of conditional probabilities on directed graphical structures. *Networks*, *20*, 579-605.

13. Olesen, K.G., Lauritzen, S.L. and Jensen, F.V. (1992). aHUGIN: A system creating adaptive causal probabilistic networks. *Proceedings of the 8th International Conference on "Uncertainty in Artificial Intelligence"*, Stanford.

14. Cowell, R.G., Dawid, A.P. and Spiegelhalter, D.J. (1993). Sequential model criticism in probabilistic expert systems. *IEEE Trans. on Pattern Analysis and Machine Intelligence*, *15*, No. 3.

15. Adams, E.N. (1984). Optimizing preventive service of software products. *IBM Journal Research and Development*, *28*, No. 1.

Single Transferable Vote: a case study of the use of VDM-SL

P. Mukherjee and B.A. Wichmann

University of Birmingham
and
National Physical Laboratory

1   Introduction

The selection of an appropriate topic for the application of formal methods is not easy. The following criteria are relevant: the example must be realistic of an actual class of applications; the example must not be too long or too specialised for those interested in the potential application of formal methods; and the 'full' details must be available which excludes many security applications and those involving commercial confidentiality.

This example has been chosen to complement the previous case study of an ammunition control system [5], since it is quite different in nature and satisfies all three of the above criteria. We must start by giving some background.

In July 1991, the Church of England changed the regulations used to specify the method of electing representatives to its main bodies. The key change was that for the first time, a computer could be used to determine the result of the election which is conducted according to the Single Transferable Vote (STV) method. These elections are controlled by statute since the Church of England is the established church. Hence, legally, the use of STV is similar to that of the Northern Ireland European Parliament public elections, except that the Northern Ireland elections do not allow the use of a computer.

It is clear, therefore, that the electorate must have absolute confidence in the use of the computer in a similar manner to that of a safety critical application (like railway signalling, for instance). Hence we believe that the application of formal methods to this example is appropriate.

1.1   1991 Synod elections

The actual regulations allowed a program to be used for Church elections provided the Electoral Reform Society (ERS) certified that the computer program does implement the specific rules that the Church uses. These rules are based upon hand counting, so in essence, the computer program must simulate the actions that would otherwise be undertaken manually.

Since the Synod elections were due in October 1991 and the regulations were only published in July 1991, only a short time was available for the certification by ERS. In fact, only about 36 man-hours could be devoted to the certification process. This was undertaken without the use of formal methods, but is reported here so that the development process can be properly understood.

The program which was eventually certified was written by I.D. Hill and consisted of about 1,000 lines of Pascal. In fact, the complete system was rather more since it included a data preparation program and other utilities to enable a complete election to be handled.

The certification was undertaken by one of the authors for the ERS and consisted of the following process.

Checking the program was based upon the analysis of statement execution applied to the main 1,000 line program. The reason for analysing just the main program was that errors elsewhere are likely to be immediately apparent, while the work needed to check the main logic is quite significant.

The main program was transferred from an IBM-PC to an Archimedes. This transfer was done for practical reasons but acted as a cross-check on the code. The program was then instrumented by hand (by using a different compiler) to discover the statements executed. Special test data was constructed to execute all the statements. All statements proved executable except two which would not be executable on either the Archimedes or IBM-PC. In fact, a program was already available to generate random test cases, so there was a potentially large source of data.

The specification of the computer program was to follow the same logic as the hand counting rules. Thus, in principle, it was easy to check the output from any specific test. However, for the larger tests, the amount of hand-checking was significant, so it was important to minimize the checking required. This was done by computing the minimum number of test cases which would ensure all the (feasible) statements were executed. This left thirteen test cases which were hand checked by the ERS expert — the ideal 'oracle' in this case.

The result of this exercise was that one significant bug was found and about six minor ones. No fault has been found subsequently in the program (although a minor fault has been found in another program concerned with the data preparation). An amusing example of a minor bug was in the handling of an election in which no votes were cast — a random choice was made (correct), but one too many people selected (wrong!).

The ability to use a computer was exercised by two dioceses in the elections to the General Synod — in fact, the elections to the Synod which made the historic decision to allow women priests.

1.2 Overview

The previous section indicates the position prior to the start of the application of formal methods to this example, namely an existing program and a set of test cases thought to ensure the correctness of at least one algorithm. The process which was followed in this case study was:

1. The formulation of the Church of England STV rules in VDM-SL, see section 2;
2. Mechanical checking of the VDM-SL using the tool SpecBox;
3. Review of the rules by hand by several people;
4. Animation of the VDM-SL using SML, see section 3;
5. Execution of the 13 test cases (devised for the certification, see above) with the SML version;
6. Revision of the VDM-SL in the light of bugs discovered during the animation.

This method is very similar to that advocated in the UK Interim Defence Standard [4]. The main difference is that there was no iteration on the specification (the ambiguities were merely noted), and that the production algorithm was produced without the aid of formal methods.

Section 4 draws some conclusions from this case study, particularly on the relevance to the production of safety-critical software to Defence Standard 00-55.

2 Specification

Below is a brief overview of the algorithm. In this section we describe the data types used in the specification and by way of example present a particular branch of the specification. We do not present formal definitions for all the auxiliary functions used here; those functions whose intended meaning is not obvious from their context will be defined informally and their formal definitions may be found in [6].

The basic algorithm is illustrated by the fragment of Ada code in Figure 1. The algorithm hinges on the concept of *the quota*. The quota is a value dependent upon the total number of valid voting papers cast and the number of vacancies to be filled. Roughly speaking, the three main ideas in the algorithm are: candidates whose total vote is greater than or equal to the quota are deemed to be elected. If, after a candidate (or candidates) has been deemed to be elected, there are still unfilled vacancies, then the candidate with the greatest vote (whether continuing or deemed to be elected), will have his or her surplus transferred (the "surplus" being the difference between the candidate's total score and the quota). If, after a candidate (or candidates) has been deemed to be elected, there are still unfilled vacancies and no candidate has a surplus, then the candidate possessing the fewest votes will be excluded and have his or her votes transferred.

```
Start;
Count_First_Preferences;
while Vacancies_Left loop
   if Number_Continuing_Candidates =Number_Unfilled_Vacancies
      then Elect_All_Continuing_Candidates;
      exit;
   end if;
   if Number_Continuing_Candidates_Satisfying_Quota <=
                              Number_Unfilled_Vacancies then
      Elect_All_Such_Candidates;
      exit;
   end if;
   if One_Vacancy and Large_Lead then
      Elect_Leading_Candidate;
   else
      if Exists_NonDeferable_Surplus then
         Transfer_Surplus;
      else
         Exclude_Candidate;
      end if;
   end if;
end loop;
```

Figure 1. Ada code illustrating basic algorithm

In this paper we describe the specification of the operations involved in choosing whose surplus to transfer. The complete specification appears in [6]. Note that for the remainder of this paper we will use the word "elected" interchangeably with the phrase "deemed to be elected".

2.1   Definitions

The three parameters of the algorithm are the number of candidates seeking election (*Number-of-candidates*), the number of vacancies to be filled (*Number-of-vacancies*) and the collection of candidate names (*Candidate-names*). The former two will not be given values in the current study. Elements of the type *Candidate-names* will be used purely to tag various different objects, so the only operation that will be performed on them is testing for equality. Hence we define the type to be the type of tokens, token. In a particular election, the names of the candidates standing is given by *Cand-names* : *Candidate-names*-set.

Below are some basic definitions; they are reproduced verbatim and are followed by some examples.

```
1. DEFINITIONS:
   -----------


In these Regulations:

(3)    The expression 'continuing candidate' means any
       candidate neither deemed elected nor excluded from the
       poll at any given time.

(4)    The expression 'first preference' means the figure '1'
       standing alone opposite the name of a candidate;
       'second' preference means the figure '2' standing alone
       opposite the name of a candidate in succession to the
       figure '1'; 'third preference' means the figure '3'
       standing alone opposite the name of a candidate in
       succession to the figure '1', '2', and so on.

(5)    The expression 'next available preference' means a
       second or subsequent preference recorded in consecutive
       numerical order for a continuing candidate, the
       preferences next in order on the voting paper for
       candidates already elected or excluded from the poll
       being ignored.

(8)    The expression 'original vote' in regard to any
       candidate means a vote derived from a voting paper on
       which a first preference is recorded for that
       candidate.

(9)    The expression 'transferred vote' in regard to any
       candidate means a vote derived from a voting paper on
       which a second or subsequent preference is recorded for
       that candidate.

(10)   The expression 'surplus' means the number of votes by
       which the total number of the votes, original and
       transferred, credited to any candidate exceed the
       quota.

(11)   The expression 'stage' means -
```

    (a)   all the operations involved in the counting of the
first preferences recorded for candidates; or

    (b)   all the operations involved in the transfer of the
surplus of an elected candidate; or

    (c)   all the operations involved in the transfer of the
votes of an excluded candidate.

Consider the voting paper shown below. For candidate Hoare, this paper
would be an original vote. On the other hand, if Hoare had been elected,
the remaining candidates were continuing and we wished to transfer this
paper from Hoare, then the next available preference would be Dijkstra
and this paper would become one of Dijkstra's transferred votes. If Hoare
had been elected, Dijkstra excluded and we wished to transfer this paper
from Hoare, then the next available preference would be Wirth.

Dijkstra	2
Hoare	1
Knuth	
Lamport	4
Wirth	3

2.1.1 *Voting-papers*

Next, we specify precisely what a voting paper shall be. From the regulations, we find the following requirements:

```
3.    VOTING PAPERS:
      --------------
```

(1)   In an election to be conducted by post, the voting
paper shall consist of a paper containing a list of
the candidates, described as their respective
nomination papers.

(2)   In an election to be conducted at a meeting, or partly
at a meeting and partly by postal voting, the voting
paper shall consist of a paper containing a list of
candidates, described in a such a way as to ensure that
the identity of each candidate is known to each voter
and to the presiding officer.

```
6.    INVALID VOTING PAPERS:
      ----------------------
```

A voting paper shall be invalid if:

```
(a)   it is not signed by a voter on its reverse; or
(b)   the figure 1 standing alone indicating a first
      preference is not placed against any candidate; or
(c)   the figure 1 standing alone indicating a first
      preference is placed opposite the name of more
      than one candidate; or
(d)   the figure 1 indicating a first preference and
      some other figure are placed opposite the name of
      the same candidate; or
(e)   it cannot be determined for which candidate the
      first preference of the voter is recorded.
```

Below are some examples of invalid voting papers. In voting paper (i), no first preference is specified (satisfying clause 6(b)). In voting paper (ii), more than one first preference is specified (satisfying clause 6(c)).

Dijkstra	2		Dijkstra	2
Hoare			Hoare	4
Knuth	5		Knuth	1
Lamport	4		Lamport	1
Wirth	3		Wirth	3

$$\qquad\qquad\text{(i)} \qquad\qquad\qquad\qquad\qquad\qquad \text{(ii)}$$

We define a voting paper to be a map from the set of candidate names to the non-zero natural numbers, each candidate name being associated with a preference.

types

$$Voting\text{-}paper = Candidate\text{-}names \xrightarrow{m} \mathbb{N}_1$$

$$\text{inv } v \triangleq \exists! \, name : Candidate\text{-}names \cdot v(name) = 1$$

Thus the following voting paper

Dijkstra	2
Hoare	4
Knuth	1
Lamport	5
Wirth	3

would be represented by

$$\{Dijkstra \mapsto 2, Hoare \mapsto 4, Knuth \mapsto 1, Lamport \mapsto 5, Wirth \mapsto 3\}$$

Note that on a physical voting paper, a name might occur with no preference next to it. In our representation of voting papers, the only names which occur are ones which have a preference next to them on the corresponding physical paper. So the following voting paper

Dijkstra	2
Hoare	4
Knuth	3
Lamport	1
Wirth	

corresponds to the map

$$\{Dijkstra \mapsto 2, Hoare \mapsto 4, Knuth \mapsto 3, Lamport \mapsto 1\}$$

We assume that all invalid voting papers are removed before transcription of papers occurs. This assumption is made explicit by the invariant that we place on voting papers. However, we clearly cannot specify clauses (a), (d) and (e).

Using this representation, it is easy to extract those names which occur (i.e., with a preference) on a voting paper by looking at the domain of the map. We can also represent voting papers in which two names might have the same preference next to them (provided that it is not the first preference, which would invalidate the paper).

2.1.2 *Transferable and non-transferable papers*

Voting papers cast may be divided into two categories: transferable and non-transferable papers:

(6)   The expression 'transferable paper' means a voting paper on which, following a first preference, a second or subsequent preference is recorded in consecutive numerical order for a continuing candidate.

(7)   The expression 'non-transferable paper' means a voting paper on which no second or subsequent preference is recorded for a continuing candidate; provided that a paper shall be deemed to have become a non-transferable paper whenever

    (a) the names of two or more candidates (whether continuing or not) are marked with the same number, and are next in order of preference; or

    (b) the name of the candidate next in order of preference (whether continuing or not) is marked -

        (i) by a number not following consecutively after some other number on the voting paper; or

(ii) by two or more numbers; or

(c) for any other reason it cannot be determined for which of the continuing candidates the next available preference of the voter is recorded.

We make an assumption here: because we have chosen to model voting papers using maps, it is not possible for us to have a voting paper where the name of a candidate is marked by two or more numbers since this would violate the well-definedness of maps. Therefore, we assume that any such papers are transcribed so that if a candidate's name is marked with two or more numbers, then that candidate and any subsequent preferences are treated as if no preference were made against them. This conforms to the regulations but places additional responsibility with the transcription process.

So for instance, the voting paper

Dijkstra	2 3
Hoare	4
Knuth	1
Lamport	5
Wirth	3

would be represented by

$$\{Hoare \mapsto 4, Knuth \mapsto 1, Lamport \mapsto 5, Wirth \mapsto 3\}$$

Note that even though there is no preference number 2, we include preferences 3,4 and 5; however, if the paper were to be transferred from Knuth then it would become non-transferable because there is no second preference.

With this proviso, most of the conditions of (7) can be translated into predicates. If *paper* represents a voting-paper and *continuing-names* represents the names of the continuing candidates, then

(7) The second and subsequent preferences on *paper* are given by

$$paper \vartriangleright \{1\}.$$

So the names of those candidates with a second or subsequent preference against them are given by dom $(paper \vartriangleright \{1\})$. Hence the first condition translates to

$$\mathsf{dom}\ (paper \vartriangleright \{1\}) \cap continuing\text{-}names = \{\ \}.$$

(a) If *discontinuing* is the name of the candidate currently credited with this paper (i.e. the current preference), subsequent preferences are given by $paper \vartriangleright \{1, \ldots, paper(discontinuing)\}$. If m is the minimum of the range of this map, then the name corresponding to m must be the next preference. So the requirement is

$$\text{card dom } paper \rhd m > 1$$

(b)(i) Using m as defined above, this translates to

$$m - 1 \notin \text{rng } paper$$

(b)(ii) With the above proviso, this case will not occur.

(c) Obviously, we cannot specify a "catch-all" clause such as this.

Combining the above predicates, we can formally characterise a paper which is non-transferable:

$$non\text{-}transferable\text{-}paper :$$
$$Voting\text{-}paper \times Candidate\text{-}names \times Candidate\text{-}names\text{-set} \to \mathbb{B}$$

$$non\text{-}transferable\text{-}paper\,(paper, disc, continuing\text{-}names) \;\triangleq$$
$$(\text{dom }(paper \rhd \{1\}) \cap continuing\text{-}names = \{\}) \;\vee$$
$$(\text{let } m = min(\text{rng }(paper \rhd \{1, \ldots, paper(disc)\})) \text{ in}$$
$$(\text{card dom }(paper \rhd \{m\}) > 1) \vee (m - 1 \notin \text{rng } paper))$$

This function is used when voting papers are transferred be it due to transfer of surplus or exclusion of a candidate.

2.1.3 *Scores and Stages*

In the process of conducting an election, we will wish to know the total number of votes currently credited to each candidate. To do this, we define the composite type *Score*, which associates with each candidate their current total value of votes credited. (Where the total value of votes is the sum over all voting papers credited to the candidate, of the value at which each paper was transferred to the candidate.)

$$Score :: name : Candidate\text{-}names$$
$$ count : \mathbb{R}$$

Note that the count is a real number rather than a natural number because when voting papers are transferred, the value at which they are transferred need not be integral.

The total value credited to each continuing or elected candidate will need to be known at the end of each stage. Therefore, we define a *Stage* to be an ordered sequence of scores of continuing and elected candidates, the ordering being on the value of votes credited to each candidate.

$$Stage = Score^{*}$$

$$\text{inv } s \triangleq \forall i, j \in \text{inds } s \cdot$$
$$((i < j) \;\Rightarrow\; (s(i).count \geq s(j).count)) \wedge$$
$$i \neq j \;\Rightarrow\; s(i).name \neq s(j).name$$

2.1.4 *Parcels, sub-parcels and bundles*

A collection of voting papers is referred to as a *Parcel*. We define a parcel
to be a map from the type *Voting-paper* to the non-zero natural numbers,
where the number associated with each paper indicates the number of occurrences of that paper.

$$Parcel = Voting\text{-}paper \xrightarrow{m} \mathbb{N}_1$$

If the papers in a parcel belonging to a particular candidate are transferred then the parcel is divided into a collection of *sub-parcels*; the papers
in each sub-parcel sharing the same next available preference. Similarly a
sub-parcel itself, when transferred, gives rise to a collection of sub-parcels.
However, when a parcel or sub-parcel is transferred, the voting papers contained therein need not necessarily be transferred at the same value that
they held originally. Therefore, each sub-parcel needs to carry the value
at which the papers contained therein were transferred. If the type *Value*
denotes the positive real numbers, then we get the following definition:

$$Sub\text{-}parcel :: votes : Parcel$$
$$value : Value$$

When a parcel or sub-parcel is transferred, the papers contained in
it are divided up as follows: all those papers which show a next available preference are divided up according to the name of the next available preference and those papers without a next available preference are
held separately and the total value of such papers recorded. Thus we
get: a collection of sub-parcels, where in any one sub-parcel, each paper
has the same next available preference, a sub-parcel containing all the non-
transferable papers and the loss of value due to the non-transferable papers.
Modeling this in VDM-SL, the last two are obviously represented using a
Sub-parcel and a real number respectively. We model the first using a map
from *Candidate-name* to *Sub-parcel*. Combining the three we get the type
Sub-parcel-bundle.

$$Sub\text{-}parcel\text{-}bundle :: sub\text{-}parcels : Candidate\text{-}names \xrightarrow{m} Sub\text{-}parcel$$
$$non\text{-}transferable : Sub\text{-}parcel$$
$$loss\text{-}of\text{-}value : \mathbb{R}$$

2.1.5 *Candidates*

During the process of electing candidates, each candidate will have a certain
number of parcels and sub-parcels containing votes with preferences for him
or her. We tag these parcels and sub-parcels with the candidate's name so
that we will be able to keep track of where all the parcels and sub-parcels
are.

The votes credited to a candidate fall into two natural categories: original votes and transferred votes. For each original vote, ov, credited to a candidate, we require the candidate is the first preference on ov and no other candidate is the first preference on ov. (In fact, assuming ov is a valid voting paper, the former implies the latter since valid voting papers have unique first preferences. However, we retain both for clarity.) For each transferred vote tv, we require the candidate's name has a preference next to it.

Again, these requirements are simple to formulate in VDM-SL. Assuming the candidate's name is $name$, the first requirement translates to the following predicate:

$$\mathrm{dom}\ ov \rhd \{1\} = \{name\}$$

The second requirement translates to:

$$\mathrm{rng}\ \{name\} \lhd ov = \{1\}$$

And the last requirement translates to:

$$name \in \mathrm{dom}\ tv$$

We therefore define a *Candidate* to be the composite type consisting of the candidate's name, the parcel of original votes and the (sequence of) sub-parcels of transferred votes.

$$
\begin{aligned}
Candidate :: \ &name : Candidate\text{-}names \\
&original\text{-}votes : Parcel \\
&transferred\text{-}votes : Sub\text{-}parcel^*
\end{aligned}
$$

$$
\begin{aligned}
\mathbf{inv}\ & mk\text{-}Candidate(name, ovs, tvs) \triangleq \\
& (\forall ov \in \mathrm{dom}\ ovs \cdot \mathrm{dom}\ ov \rhd \{1\} = \{name\} \land \\
& \qquad\quad \mathrm{rng}\ \{name\} \lhd ov = \{1\}) \land \\
& (\forall sub\text{-}parcel \in \mathrm{elems}\ tvs \cdot \\
& \qquad \forall tv \in \mathrm{dom}\ sub\text{-}parcel.votes \cdot name \in \mathrm{dom}\ tv)
\end{aligned}
$$

We are required to use a sequence for the collection of sub-parcels because of the following regulation:

```
12.   PAPERS TRANSFERRED TO BE PLACED ON TOP OF PARCEL
      ------------------------------------------------------


Whenever any transfer is made under any of the preceding
Regulations each sub-parcel of papers transferred shall be
placed on top of the parcel or sub-parcel, if any, of papers
of the candidates to whom the transfer is made, and that
candidate shall be credited with a value ascertained in
pursuance of these Regulations.
```

2.1.6 *State*

At the end of any stage in the electoral process, to be able to conduct the next stage, we will require the sets of continuing, excluded and elected candidates, the sequence of stages that have already occurred (in chronological order) and the quota.

Note that the sets of continuing, excluded and elected candidates must be disjoint and the names of the candidates therein must span the type *Candidate-names*. Also, we require that a candidate is uniquely defined by name.

We encapsulate the above information in the state of the specification and specify the above requirements in the invariant on the state.

> state *St* of
>   *elected* : *Candidate*-set
>   *excluded* : *Candidate*-set
>   *continuing* : *Candidate*-set
>   *stages* : *Stage**
>   *quota* : $\mathbb{R}$
>
>   inv *mk-St*$(el, ex.co, s, q)$ $\triangleq$
> $(\{cand.name \mid cand \in el \cup ex \cup co\} = Cand\text{-}names) \wedge$
> $disjoint(\{el, ex, co\}) \wedge$
> $(\forall\, cand_1, cand_2 \in el \cup ex \cup co \cdot$
> $cand_1 = cand_2 \Leftrightarrow cand_1.name = cand_2.name)$
>
> end

2.2 Beginning an election

Once all voting papers have been cast, there are three main operations involved in performing the first stage of the election: arrange voting papers into parcels according to first preference then credit each candidate with the number of first preferences counted; on the basis of the number of voting papers cast and the number of candidates, calculate the quota; and elect any candidates whose value is greater or equal to the quota, provided the number of such candidates does not exceed the number of vacancies.

We shall describe the first two operations here but leave the description of the third until section 2.2.3.

2.2.1 *Sorting and counting papers*

From the regulations, we find the following information:

```
7.    SORTING AND COUNTING PAPERS:
      ----------------------------


(1)   The presiding officer, after rejecting any voting
      papers that are invalid, shall cause the valid voting
```

```
papers to be arranged in parcels according to the
first preferences recorded thereon for each candidate.
```

(2)   The presiding officer shall count the number of papers
        in each parcel, and credit each candidate with a number
        of votes equal to the number of valid papers on which a
        first preference has been recorded for such candidates.

As explained in Section 2.1.5, a vote v is a first preference for candidate $name$ if dom $v \rhd \{1\} = \{name\}$. Hence, given a parcel of votes $votes$, the collection of first preferences for $name$, will be the sub-map of $votes$ whose domain satisfies the above condition, namely

$$\{v \mid v \in \text{dom } votes \cdot \text{dom } v \rhd \{1\} = \{name\}\} \lhd votes$$

Clearly, in the first instance no candidate will have any transferred votes. Thus we get the VDM-SL function $sort\text{-}papers$ which distributes voting papers according to first preference.

$$sort\text{-}papers : Parcel \times Candidate\text{-}names\text{-set} \to Candidate\text{-set}$$

$$
\begin{aligned}
&sort\text{-}papers\,(votes, names) \triangleq \\
&\quad \{mk\text{-}Candidate(name, orig\text{-}votes, [\,]) \mid \\
&\qquad name \in names, orig\text{-}votes : Parcel \cdot \\
&\qquad\quad orig\text{-}votes = \{v \mid v \in \text{dom } votes \cdot \\
&\qquad\qquad\qquad \text{dom } v \rhd \{1\} = \{name\}\} \lhd votes\}
\end{aligned}
$$

To satisfy requirement (2) we need to count the number of votes each candidate has received. We do this using the function $size$, defined in [6], which counts the number of voting papers in a parcel, then we build a stage. Using the invariant of the type $Stage$, we only need to describe the elements of the stage; the remaining structure is implied by the type invariant. Thus we get the function $build\text{-}first\text{-}stage$.

$$
\begin{aligned}
&build\text{-}first\text{-}stage\,(candidates : Candidate\text{-set})\ stage : Stage \\
&\text{post elems } stage = \\
&\quad \{mk\text{-}Score(candidate.name, size(candidate.original\text{-}votes)) \mid \\
&\qquad\qquad\qquad candidate \in candidates\} \land \\
&\quad \text{len } stage = \text{card elems } stage
\end{aligned}
$$

2.2.2 *Calculating the quota*

The *quota* is the value of voting papers which any candidate must satisfy to be elected. The formula for calculating the quota is given by the following regulation:

```
8.    QUOTA:
      -----


The presiding officer shall add together the number of votes
credited to all the candidates and then divide the sum by a
number exceeding by one the number of vacancies to be filled,
the division being continued to two decimal places. If the
result is not exact the remainder after two decimal places
shall be disregarded, and the result increased by 0.01. This
number, being sufficient to ensure the election of a
candidate, is herein called the 'quota'.
```

That is, the quota is given by:

$$quota = \frac{total\ number\ of\ valid\ papers\ cast}{1 + number\ of\ vacancies\ to\ be\ filled}$$

with the value being rounded up after two decimal places if it is not exact to two decimal places. We say that a candidate "satisfies" the quota if the value of votes credited to that candidate is no less than the quota.

2.2.3 *Elected candidates*

We now define precisely what it means for a candidate to be deemed elected. If after the initial count or after a surplus is transferred, one or more continuing candidates satisfy the quota, then those candidates are elected, subject to the following two regulations:

```
9.    CANDIDATE WITH QUOTA DEEMED TO BE ELECTED:
      ---------------------------------------------


If the value credited to a candidate is equal to or greater
than the quota, that candidate shall be deemed to be elected,
provided that the number of candidates deemed to be elected
does not exceed the number of vacancies to be filled.


10.   SURPLUS TO BE TRANSFERRED:
      --------------------------

(8)   Candidate deemed to be elected after transfer of a
      surplus:

      If, after the transfer of a surplus, the value
```

> credited to a continuing candidate is now equal to
> or greater than the quota, that candidate shall be
> deemed to be elected, provided that the number of
> candidates deemed to be elected does not exceed the
> number of vacancies to be filled.

Assuming the auxiliary functions *num-candidates-satisfying-quota* and *numbe of-remaining-vacancies* which respectively return the number of candidates with a total value of votes exceeding the quota and the number of vacancies still to be filled (details of which may be found in [6]), this leads directly to the following operation:

CHANGE-STATUS-OF-ELECTED-CANDS ()
  ext rd *stages* : *Stage**
    rd *quota* : $\mathbb{R}$
    wr *continuing, elected* : *Candidate*-set
  pre *num-candidates-satisfying-quota*(*continuing, stages, quota*)
        $\leq$ *number-of-remaining-vacancies*(*elected*)
  post let *candidates-satisfying-quota* =
      {*candidate* | *candidate* $\in \overleftarrow{continuing}$ ·
        $\exists$ *count* : $\mathbb{R}$ ·
            (*mk-Score*(*candidate.name, count*) $\in$ elems hd *stages*)
            $\wedge$ (*count* $\geq$ *quota*)} in
    (*elected* = *candidates-satisfying-quota* $\cup \overleftarrow{elected}$) $\wedge$
    ($\overleftarrow{continuing}$ = *continuing* $\cup$ *candidates-satisfying-quota*)

Note that candidates who satisfy the quota after having received the votes of an excluded candidate are elected separately.

2.2.4   *Initialisation*

Here we describe the operation which initialises the state. We begin by counting voting papers and sorting them according to candidates, to get the local value, *cur-cont*. Candidates cannot be excluded by this operation so we initialise the excluded set to be empty. If any candidate satisfies the quota, provided that the total number of such candidates does not exceed the number of candidates to be elected, such a candidate will be deemed to be elected, using the predicate *post-CHANGE-STATUS-OF-ELECTED-CANDS* and the set of continuing candidates will be all other candidates. Otherwise, we set *elected* to be empty and *continuing* to be *cur-cont*. The quota will be as described in 2.2.2 and the stage will be constructed using the function *build-first-stage*. So we get:

PREPARE-ELECTION (*votes* : *Parcel*)
ext wr *excluded, elected, continuing* : *Candidate*-set
   wr *stages* : *Stage**
   wr *quota* : $\mathbb{R}$
post (*excluded* = {}) $\wedge$
    let *cur-cont* = *sort-papers*(*votes, Cand-names*) in
    (*quota* =
      two-dec-places(*size*(*votes*)/(*Number-of-vacancies* + 1))) $\wedge$
    (*stages* = *build-first-stage*(*cur-cont*)) $\wedge$
    if 0 < *num-candidates-satisfying-quota*(*cur-cont, stages, quota*)
      $\wedge$ *num-candidates-satisfying-quota*(*cur-cont, stages, quota*)
         $\leq$ *Number-of-vacancies*
    then *post-CHANGE-STATUS-OF-ELECTED-CANDS*(
         mk-St({ }, { }, *cur-cont, stages, quota*),
           mk-St(*elected*, { }, *continuing, stages, quota*))
    else (*elected* = { } $\wedge$ *continuing* = *cur-cont*)

where *two-dec-places* is the auxiliary function which rounds a real number
to two decimal places.

2.3 Transfer of surplus

We now come to consider how and when a candidate having a surplus shall
have his or her votes transferred. All regulations quoted in this section will
be taken from section 10 of the regulations.

2.3.1 *Choosing which surplus to transfer*

The following regulation tells us when a candidate's surplus should be
transferred.

```
10.   SURPLUS TO BE TRANSFERRED:
      -------------------------
```

(1)   Value greater than quota: surplus transferred:

     If at the end of any stage of the count the value
     credited to one or more candidates is greater than the
     quota, the presiding officer shall, subject to the
     provisions of this Regulation, transfer the largest
     surplus. If two or more candidates each have an equal
     surplus, the presiding officer shall transfer the
     surplus of the candidate who was credited with the
     greatest value at the earliest stage at which the
     values credited to such candidates were unequal. If the
     values credited to such candidates were equal at all
     stages of the count, the presiding officer shall

Name	Value
Hoare	340
Dijkstra	160
Knuth	150
Wirth	140
Lamport	130
Milner	100

(i)

Name	Value
Dijkstra	210
Hoare	170
Wirth	190
Knuth	180
Lamport	150
Milner	120

(ii)

Name	Value
Wirth	200
Knuth	200
Dijkstra	170
Hoare	170
Lamport	150
Milner	130

(iii)

Figure 2. Example of transfer

```
determine by lot which surplus he shall transfer.
```

To clarify this regulation, we give an example. We wish to elect five people from the six candidates

$$\{Dijkstra, Hoare, Knuth, Lamport, Milner, Wirth\}$$

1020 valid voting papers were cast, giving a quota of 170. After the voting papers were distributed according to first preference and the number of papers credited to each candidate counted, the value of votes credited to each candidate was as shown in stage (i) of Figure 2.

Here, Hoare is the only candidate with a surplus, so his surplus of 170 gets transferred. After the transfer we get the situation shown in stage (ii) of 2.

Now Dijkstra has a surplus of 40 which is transferred. Note that now Wirth has overtaken Knuth.

Following the transfer of Dijkstra's surplus we get stage (iii) of Figure 2. Both Wirth and Knuth have surpluses to be transferred but they have the same value. However, because at the earliest stage at which they were not equal Knuth had a greater score than Wirth, Wirth has his surplus transferred.

This process would continue until five candidates were elected or no candidate had a transferable surplus, in which case the lowest placed candidate would be excluded.

We now translate these ideas into VDM-SL. We begin by defining the predicate *sole-leader* which takes a stage (*stage*), a candidate's name (*name*) and a set of candidate's names (*leaders*) and is true exactly when *name* has a value of votes credited which is strictly greater than that of any member of *leaders* at this particular stage.

sole-leader :
$\qquad$ *Stage* $\times$ *Candidate-names* $\times$ *Candidate-names*-set $\rightarrow$ $\mathbb{B}$

sole-leader (*stage, name, leaders*) $\triangleq$
$\quad$ let *cand* $= \iota\, c \in$ elems *stage* $\cdot$ *c.name* $=$ *name* in
$\quad$ let *leading-scores* $=$
$\qquad\qquad$ $\{sc \mid sc \in$ elems *stage* $\cdot$ *sc.name* $\in$ *leaders*$\} \setminus \{cand\}$ in
$\quad$ $\forall\, sc \in$ *leading-scores* $\cdot$ *cand.count* $>$ *sc.count*
$\quad$ pre $\;\exists\, c \in$ elems *stage* $\cdot$ *c.name* $=$ *name*

Using *sole-leader*, we can define the predicate *greatest-value-at-earliest-stage*. This takes a candidate's name (*name*) and all the stages that have occurred so far (*all-stages*) and is true precisely when there exists a stage at which *name* satisfies *sole-leader* and no other candidate with a surplus currently equal to *name*'s surplus satisfies *sole-leader* at any earlier stage.

greatest-value-at-earliest-stage : *Candidate-names* $\times$ *Stage** $\rightarrow$ $\mathbb{B}$

greatest-value-at-earliest-stage (*name, all-stages*) $\triangleq$
$\quad$ let *leaders* $= \{$*score.name* $\mid$ *score* $\in$ elems hd *all-stages* $\cdot$
$\qquad\qquad\qquad\qquad$ *score.count* $=$ hd hd *all-stages.count*$\}$ in
$\quad$ $\exists\, i \in$ inds *all-stages* $\cdot$
$\qquad$ *sole-leader*(*all-stages*(i), *name, leaders*) $\wedge$
$\qquad$ $(\forall\, j \in \{i+1, \ldots, $ len *all-stages*$\},$ *other-leader* $\in$ *leaders* $\cdot$
$\qquad\qquad$ $\neg$ (*sole-leader*(*all-stages*(j), *other-leader, leaders*)))

Now, choosing which surplus to transfer is trivial. We do this as follows:

1. Extract the names of those candidates whose count equals the current maximum count. Call this set *leaders*.
2. If *leaders* is singleton, then no more work need be done.
3. If *leaders* contains more than one element, we get

   Either (a) At some earlier stage, a member of *leaders* had a vote strictly greater than all the other members of *leaders* – in which case *greatest-value-at-earliest-stage* is satisfiable.

   Or (b) The members of *leaders* had the same score as each other at every previous stage, in which case we return a random element of *leaders*.

This can be expressed concisely in VDM-SL:

CHOOSE-SURPLUS-TO-TRANSFER () *name*: *Candidate-names*
ext rd *stages* : *Stage**
pre *stages* $\neq$ []

$$\text{post let } leaders = \{score.name \mid score \in \text{elems hd } stages \cdot$$
$$score.count = \text{hd hd } stages.count\} \text{ in}$$
$$(name \in leaders) \land$$
$$((\text{card } leaders > 1) \Rightarrow$$
$$(\text{if } \exists\, n \in leaders \cdot greatest\text{-}value\text{-}at\text{-}earliest\text{-}stage(n, stages)$$
$$\text{then } greatest\text{-}value\text{-}at\text{-}earliest\text{-}stage(name, stages)$$
$$\text{else } post\text{-}RANDOM\text{-}ELEMENT(leaders,$$
$$mk\text{-}St(\{\,\}, \{\,\}, \{\,\}, [], 0), mk\text{-}St(\{\,\}, \{\,\}, \{\,\}, [], 0), name)))$$

3 Animation

3.1 The principles

Given a specification of reasonable complexity, it is clear that it is very hard to be totally confident of its correctness. The last section indicates that in this case, the complexity of the STV algorithm is quite high, in spite of an English description of modest length.

The method chosen here to validate the VDM-SL specification is one of *animation*. By this, we mean converting the VDM-SL into a form which can be directly executed, and using the execution as a means of checking the correct operation of the specification. The conversion of an arbitrary VDM-SL specification into an executable form is not, in general, feasible, since VDM-SL allows the use of implicit specifications. Hence if an algorithm is specified by a post-condition, then there may be no way of providing an implementation, and indeed, the post-condition could prove impossible to satisfy. However, one objective of the animation exercise would be to highlight such unsatisfiable post-conditions.

In the current study, the VDM-SL specification is algorithmic and therefore the problem with implicit specifications does not arise. However, the conversion must follow the VDM-SL precisely, otherwise the validation process will be flawed. In consequence, the automatic conversion of the VDM-SL would be advantageous, or failing that, the conversion to a language which has similar constructs to those used in the VDM-SL so that the conversion process itself can be easily checked.

Given an executable version of the specification, it is essential to execute it in as searching a manner as possible if confidence is to be gained in the specification. In this case, we have an almost ideal situation of 13 carefully constructed test cases which are thought to be adequate to validate (another version) of the algorithm.

3.2 SML

SML originally arose as a tactics language for use in theorem proving in LCF logic [3]. However, it has developed into a respectable programming language in its own right. SML has functional programming features such as lists, recursive data types and curried functions as well as imperative features such as assignment, sequential composition and loops. This latter feature is based on the presence of "reference variables".

The presence of both of these facilities makes SML an attractive choice to animate VDM-SL specifications. This is for four reasons:

- The presence of lists in SML (which are dynamic structures) provides a natural way to interpret sets, sequences and maps. In other languages we might have to build separate abstract data type modules for these to cope with their changing size during computation.
- Functions in VDM-SL can be interpreted directly as functions in SML.
- Operations in VDM-SL can be implemented as SML functions over reference variables.
- Operation pre-conditions can be checked in SML using SML's exception handling mechanism.

Few languages satisfy the four requirements listed above other than SML, so SML was an obvious choice.

3.3 Example

To illustrate how animation was performed, we describe the SML version of the function *sole-leader* (the VDM-SL definition of which is given on page 200). Since all sets in VDM-SL are finite, they were implemented using SML's lists, therefore set comprehensions were translated into list expressions using the functions `map` and `filter`. Also, universal quantification was implemented using distributed conjunction over a list (denoted by the SML operator `conjl`). Records in VDM-SL were implemented using SML's records, the difference being in the syntax of selection, e.g., if c is a candidate then in VDM-SL the name of c would be given by $c.name$ whereas in SML the name of c would be given by `#name(c)`. Thus the SML version of *sole-leader* was:

```
fun sole_leader (stage:Stage, name:Candidate_names,
                 leaders:Candidate_names set) =
  let val name_sc = hd (filter (eq name o (#name)) stage)
      val leading_scores =
                filter ((mem (set_seq leaders)) o #name) stage
      fun gt (sc1:Score) (sc2:Score) = sc1 = sc2 orelse
                            (#count(sc1)) > (#count(sc2))
  in
    conjl (map (gt name_sc) leading_scores)
  end;
```

3.4　Errors discovered by animation process

Testing of the SML program was performed against a suite of test data first used for testing the original Pascal program for performing STV elections. 13 test cases were used for the original program but we only used 10 of these tests here because our program did not implement options such as withdrawal of a candidate following casting of votes. We do not describe the data sets – details of these may be found in [6].

Before animation the specification was analysed using the tool SpecBox [1]. SpecBox checks VDM-SL specifications for conformance with BSI-VDM syntax and performs weak type analysis. The type analysis it performs is based on arity checking: function and operation calls are checked to make sure that the number of arguments provided conforms to the number of parameters in the corresponding definitions. In fact, SpecBox was used throughout the construction of the specification and uncovered many syntax and type errors. However, as described below, a few type errors remained.

Animation of the specification revealed the following errors: two type errors in the specification (i.e. type errors missed by SpecBox); three errors due to correct interpretation but incorrect specification of requirements; two errors due to misinterpretation of English requirements; and one error introduced during translation of VDM-SL into SML. A more detailed discussion of the errors uncovered by animation may be found in [6].

4　Discussion

4.1　Was VDM-SL suitable?

One can argue that VDM-SL is being used as a programming language in this case study, and therefore perhaps a different choice should have been made. Using a conventional programming language instead would have avoided the conversion process for animation, so the balance would seem to be against VDM-SL. However, a conventional programming language, even Pascal, has quite complex semantics which could make the code difficult to understand. In fact, some of the limitations of Pascal, particularly that functions cannot return composite values, mean that a version of the STV algorithm in Pascal would be very different from the VDM-SL.

In fact, the Pascal program which was certified, contains a significant fraction of the code for the input and output which is not formally defined in section 2. This is an inherent danger in the use of a conventional language in that implementation considerations tend to predominate so that the simple expression of the semantics is lost amongst other verbiage.

In this case, we feel that the use of VDM-SL *has* proved its worth. The semantics of the algorithm has been captured at the suitable level of abstraction. The complexity of the VDM-SL reflects that of the STV algorithm. Part of the difficulty in reading the algorithm is that the notation of sets and maps is less familiar than, say, differential calculus (even though the underlying mathematics is simpler).

4.2 The value of animation

The use of animation in this case was highly effective, due to the simple conversion process and the existence of good test data. Note that we cannot claim that the VDM-SL specification is necessarily correct because it has passed the validation process. Indeed, the once-through process revealed eight errors overall, so that one could argue that at least one further error might be found via extensive back to back testing of the SML and Pascal versions.

The most worrying aspect of the process is that the number of bugs per source line of VDM-SL is rather similar to that noted from conventional programs. This implies that reliance upon a formal specification without any animation (or other validation procedure of similar strength) seems unwise.

4.3 The 00-55 development model

Three parts of the Interim Defence standard [4] are particularly relevant to this study and therefore we repeat them in full:

> "29.3.1 The Design Team shall validate the Software Specification against the Software Requirements Specification by animation of the Formal Specification. Animation shall be carried out by both of the following:
>
> - By the construction of Formal Arguments in accordance with 32.1 showing that the Formal Specification embodies all the safety features described in the Software Requirements Specification.
> - By the production of an Executable Prototype derived from the Formal Specification. The minimum number of changes shall be made to the Formal Specification in the construction of the Executable Prototype; Formal Arguments shall be constructed in accordance with 32.1 to show the link between the two.
>
> 29.3.2 The Executable Prototype shall be tested by the Design Team. The aim of each test and the expected results shall be determined before the test is executed. As a minimum, the tests shall exercise all the safety functions of the SCS. The results shall be recorded in the Software Specification.
>
> 33.3 The integrated SCS shall be tested in a test harness that enables the results to be compared with the Executable Prototype. During these tests, the SCS shall be run on the target hardware or an emulator that has been shown by Formal Ar-

guments to be equivalent to the target hardware."

In this study, we do not have safety features, and hence the first itemised paragraph cannot be undertaken as it stands. However, the general requirement that if there are n places, then exactly n people must be elected should be capable of formal verification. Indeed we prove this in [6].

We believe that we have followed the spirit of the second itemised paragraph. The executable prototype followed the specification very closely with the exception of the additional input-output which is obviously necessary. Prototyping tools should perhaps provide a simple means of undertaking input-output to simplify the process. The link between the SML code and the VDM-SL has not been subject to Formal Arguments in the sense of the standard. We do not think this would be worthwhile, bearing in mind the close correspondence between the two.

Concerning section 29.3.2, we believe the testing is adequate. The fact that the results were checked by the ERS expert, and that the algorithm does mimic the manual method, gives us a high degree of confidence in the validation of the SML code.

On the question of section 33.3, the position is not quite so strong. No attempt has been made to use the 'target hardware' (the Church program uses an IBM-PC). Indeed, configuration problems can arise, such as lack of disc space and conflicts in file names, etc, unless elaborate checks are undertaken. It would be possible to undertake back-to-back testing of the Pascal program with the SML version. However, there are some practical problems with this. Firstly, the output would have to be identical if automatic comparison of the output was to be used. Secondly, in the case of ties, the computer programs can legally make different choices, and hence some means must be devised of avoiding this if completely automatic comparisons were to be undertaken. Of course, it would be feasible to undertake further test cases with manual checks on the output, but at this stage, any remaining errors are very unlikely, and hence this option is not attractive.

4.4   Conclusions

We believe that this case study supports the use of Formal Methods. Ambiguities were located in the English text, although the experts (in ERS) have an agreed interpretation of these points. The problems of specifying an algorithm in English are well-known [2]. Mistakes do occur in public elections and one of the attractions of using computers would be to reduce the frequency of such errors. However, errors could not be eliminated completely by such means, even if the program were error-free, since the input of the data is likely to be a significant source of errors.

The results obtained here strongly support the need for validation of a formal specification, and that providing an executable prototype would appear to be a good method for this, provided the execution tests are thorough enough.

We have not undertaken back-to-back execution of the executable prototype and the production program. This would appear to be a strong method of testing, provided the comparison process can be automated. It would be easy to envisage situations in which such an automatic comparison would be difficult. For instance, if the specification involves real arithmetic, there may be different rounding between the two systems, given differences which are nevertheless within the specification. In general, implicit specification methods can easily yield different but valid implementations. Such a situation could make back-to-back testing uneconomic due to the need to undertake large numbers of hand comparisons or write elaborate software for an automatic process.

It must be admitted that the use of a formal notation for the expression of the STV algorithm is much less attractive than that produced for the formulation of the regulations for the storage of explosives [5]. In general, one must expect the formal specification to reflect the closeness of the subject matter to classical mathematics. The regulations for the storage of explosives can be represented as a set of pure functions, and hence the main complexity is in the classification of the explosives themselves and in the buildings (which is easier to understand). It is difficult to obtain the same level of understanding of the STV algorithm without conducting some elections by hand.

It did not prove possible to validate the explosives example by animation, since searching test cases were not available. This study therefore provides better insight into animation. This indicates that one should expect faults in specifications to appear at a comparable frequency to that of programs, say one fault every 70 lines of VDM-SL, which was the rate observed here.

Acknowledgements

We would like to thank I.D. Hill for pointing out some errors in an early version of the specification. One of the authors was supported by a CASE award jointly funded by the National Physical Laboratory and the Science and Engineering Research Council.

Bibliography

1. Froome, P.K.D. (1990). *SpecBox*, Adelard Software.
2. Hill, I.D. (1972). Wouldn't it be nice if we could write computer programs in ordinary English – or would it? *The Computer Bulletin, 16.*
3. MacQueen, D., Harper, R. and Milner, R. (1986). Standard ML. Technical Report ECS-LFCS-86-2, Department of Computer Science, University of Edinburgh.

4. MOD (1991). The Procurement of Safety Critical Software In Defence Equipment. Interim Standard 00-55 Issue 1, Ministry of Defence, Directorate of Standardization, Kentigern House, 65 Brown Street, Glasgow G2 8EX.

5. Mukherjee, P. and Stavridou, V. (1993). The formal specification of safety requirements for storing explosives. *Formal Aspects of Computing*, 5, 299–336.

6. Mukherjee, P. and Wichmann, B.A. (1993). STV: A case study in VDM. Technical Report DITC 219/93, National Physical Laboratory.

Mathematics of computation for (Software and other) Engineers

D.L. Parnas

Communications Research Laboratory, Department of Electrical and Computer Engineering, McMaster University, Canada

1 Preliminary provocation

The title of this paper implies that Software Engineers are Engineers, i.e. that "Software" plays the same role in their title that "Electrical", "Mechanical", or "Chemical" play in the titles of other engineering specialities. This, in itself, would seem to be a controversial statement, since it suggests that the model for software engineering education should be engineering education, not science education or mathematical education. That is my opinion, and one of the assumptions underlying this paper, but it is not the subject of this paper.

I like the term, "mathematical engineer", which I am told is used by some Dutch Technical Universities for software engineering. It seems to me that, just as certain areas of electrical physics comprise the basic knowledge of an Electrical Engineer, certain areas of mathematics, which comprise (in my opinion) the most substantive areas of computer science, should be the basic knowledge that characterises software engineering. However, we should not forget that just as Chemical Engineers need to know much more than chemistry, Software Engineers will need to know more than Computer Science and Mathematics. There must be some fundamental areas of Computer Science and Mathematics that we don't have time to teach.

2 The role of mathematics in engineering

Those who do not have an engineering education are often not aware just how much mathematics is taught to engineers. At my university, approximately 30% of an engineer's education is devoted to things that are explicitly titled mathematics. There is also a great deal of mathematics taught in specialised engineering courses. This is not atypical; it is often required by accreditation committees, which control whether or not the graduates of a programme can easily gain recognition as professional engineers.

Ability to use mathematics is one of the things that differentiate professional engineers from technicians. A major emphasis in engineering education is the concept of professional responsibility. An Engineer is taught from her first day at University, that her products must be "fit for use". Engineering students learn that they cannot trust their intuition to be sure that a product is "fit for use". Much of their education is devoted to learning how to perform *both* mathematical analysis, and carefully planned testing, of their products. They also learn to accept, as completely normal, the fact that their work will be subject to careful analysis and criticism, often based on mathematical analysis, by others. My own engineering education included approximately as much mathematics as would have been taken by a mathematics major and, at my *alma mater*, many of the courses were taken together with the mathematics majors. Regrettably, it is increasingly common to find special engineering mathematics courses, and to find that the mathematics professors who teach those courses assume that they are teaching people whose intellectual level is not as high as that of mathematicians. Having taught both engineers and mathematics majors, I do not see differences in ability, but I do see differences between the viewpoint of engineering students, and those of students majoring in mathematics or science.

Although engineers study a lot of mathematics, an engineer's view of mathematics is substantially different from that of mathematicians. Roughly put, engineers can take a lot for granted. Because they use mathematics for the description and analysis of some physical product, they implicitly assume that functions have the properties that all functions describing physical products must have. Engineers use mathematics to talk about real things; because they know what "things" they are talking about, they often do not appreciate some of the philosophical issues that arise when mathematicians try to work with abstract concepts without specific concrete applications. They often do not bother to state their assumptions explicitly. This appears sloppy to many "formalists". In most cases, the mathematics is perfectly sound provided that one makes the implicit assumptions explicit in an environmental declaration. Because engineers are working in situations where it is clear which symbols in their equations are variables, and they know what they are trying to compute, they see little need for the quantifiers, type and signature declarations that logicians demand of their colleagues. Whereas mathematicians are primarily interested in deep theorems and general properties of classes of expressions, engineers are often concerned with "junk" theorems and detailed analyses of special cases. In such situations the careful habits of logicians seem quite impractical and there is always a gap between a logician's treatment of a subject and that of an engineer who uses the same fundamental mathematics. What the logician chooses to record explicitly, the engineer often assumes without much discussion. Those interested in exploring such issues further, should look at some of the writings of N.G. de Bruijn and his students; they had to pay a lot of attention to the "short-cuts" used by working mathematicians and engineers when they were developing their "Automath" system [6].

It must also be recognised that mathematics is often implicit, rather than explicit, in engineering notations. When an electrical engineer notes the inductance, resistance, and capacitance of a component, she knows that these are the parameters for a set of differential equations. Those equations need not be written down; they are just used when necessary. Again, we see that engineering notations assume things that a mathematician would state explicitly.

These remarks lead me to a pair of preliminary conclusions:

• Engineers, whether software or otherwise, can be expected to make extensive use of mathematics in the analysis of their products, including programs. Those who refuse to do so, are technicians, not engineers.

• When we develop mathematical methods for use by engineers, we need to respect the traditional differences between engineering mathematics and the type of mathematics promoted by "formalists" or logicians in the style of Hilbert. If we don't, we will be unnecessarily frustrated and quite ineffective.

Although engineers need a broad background in mathematics, this paper focuses on a small area, the mathematics that can be considered the "science of programming". Other mathematical topics belong in the "knowledge base" of engineers, but will not be discussed here. For example, every engineer, including software engineers, requires a knowledge of probability and statistics, but that will not be discussed further in this paper.

3 The role of programming in engineering

When I was an undergraduate, programming courses were optional. Moreover, no academic credit was given for them. The computer was considered to be a slightly enhanced version of the mechanical calculator. There was no more thought of including a computer course in the curriculum than we would think of including a course on the Marchand calculators that filled some laboratories, or a course on the slide-rules that many of us carried on our belts. It was expected that we would learn to use these "tools of our trade" on our own, or in non-credit courses. Programming was considered to be a simple mechanical task, "laying down instructions", akin to wiring up a circuit. Many engineers had never taken a course in programming. When we began to offer the first credit-bearing course in programming, there were many who feared that it would not have intellectual content analogous to a physics or calculus course. The Computer Science Department had to promise that they would not simply teach a specific programming language, but would teach something deeper and more lasting.

Today, things have changed - both for the better and for the worse. There is no longer any question about whether or not an engineer should have courses in programming. Computers and software are now ubiquitous in engineering. Many engineering products include computers and software; many others are designed and analysed using computers. Hardly a week passes in which we do not hear some anecdote about the failure of an engineering product being caused either by the software contained in it or by an error in the software used to design it. Since people rarely talk loudly about their failures, we can assume that the anecdotes are just the "tip of the iceberg". Nobody questions the need for engineers to be good programmers and good at evaluating the software that they use.

However, there is something else that few question: many do not question the intellectual content of many engineering courses in computing. Nobody asks whether the intellectual content of these courses is comparable to that of other math or science courses. Perhaps the question is not asked because the answer would be embarrassing. The typical programming course simply teaches a programming language, an artifact designed by a few fallible human beings. Most of the time is spent on things that are not mathematical truths, or facts about the universe; they are just design decisions by (often not very good) language designers. The courses are exactly equivalent to teaching about a particular calculator, including the location of its buttons, how to turn it on, how to change the display, etc. Many of these courses teach almost the same artifacts that were taught 30 years ago, but that is not the real problem. The real problem is that the subject of the course is the artifact. You can always tell that something is wrong when there is a big debate about which artifact to teach. The situation is analogous to changing the lectures of a course on electrical circuit theory because we acquired new oscilloscopes. Another sure sign that something is wrong comes when someone defends a course by saying that they just introduced a new artifact.

We must also recognise another difference between engineering education and the education of scientists and mathematicians. In engineering schools there is a major emphasis on design. We are required by our accreditation committees to identify a large part of our curriculum as design. Design and analysis can be understood as complementary skills. Design is inherently creative and all that we can teach are heuristics, things that don't always work; consequently, solid, disciplined analysis is necessary. Mathematics is taught as part of the analysis component of these courses. This is in sharp contrast to the attitudes taken by another famous Dutchman in our field. E.W. Dijkstra, and many others, like to talk about deriving programs from specifications. This is not the attitude taken in other areas of engineering. Design is recognised as a very creative task, in which mathematics and science provide essential inputs; the primary role of the mathematics comes in the documentation and validation of the design. Program derivation from requirements appears analogous to deriving a bridge from a description of the river and the expected traffic. Refining a formal specification into a program would appear to be like refining a blueprint to produce a house. Engineers always make a distinction between the product itself and a description of it. This distinction seems to have lost in the computer science literature on programming. Mathematical tradition, in which the symbols are the product, dominates.

Those who chose engineering as a career path are often people with a fairly pragmatic view of life. They appreciate mathematics that is simple and elegant but they want frequent assurance that the mathematics is useful. It is important to show them how to use a mathematical concept, not simply to teach them the definitions and theorems. In engineering mathematics the emphasis has always been more on application of theorems than on proofs.

4 The mathematics needed for professional programming

I have recently taught a new course for first year engineers of all specialities. It replaced a course that could have been taught 30 years ago. I made two major changes:

- A large part of the course taught the basic mathematics behind programming with emphasis on the use of mathematics to describe what a program does, or must do, without giving an algorithm. All programming assignments were expressed as mathematical specifications.

- It was made very clear that the programming language was not the subject of the course. Students were given a choice of two programming languages that could be used in the laboratories. Two of the three lectures per week were taught in an algorithmic notation based on Dijkstra's guarded commands. The third, "laboratory", lecture taught a "real" language.

The course emphasised both the creative steps in programming and the analytical steps needed to confirm that one had not just created a monster.

The remainder of this section describes the mathematical contents of that course and how we used the mathematics to teach programming.

4.1   Finite state machines

The first step in getting students to take a professional approach to programming is to get rid of the "giant brain" and "obedient servant" views of a computer. It is essential that students see computers as purely mechanical devices, capable of mathematical description. Students are taught that "remembering" or "storing" data is just a state-change, and taught to analyse simple finite state machines to "show" that they accomplish simple recognising tasks. The Moore-Mealy model is used.

4.2   Sets, functions, relations, composition

We present the basics of a naive set theory in which all sets consist of a finite number of elements selected from previously defined universes. We present the concept of relations (functions) as sets, and the operation of union, intersection, negation and functional (relational) composition. It is important to present the students with examples of the use of these concepts and exercises in their use. We want the students to know far more than the definitions and the algebraic laws; we ask them to apply the concepts to provide precise models of real-world situations. We show how the state machines that they learned about can be described by a pair of mathematical relations.

4.3   Mathematical logic based on finite sets

In the first two sections of the course, finite state machines and sets are kept not just finite, but small, so that they can all be described by enumeration. The next step is to point out that these are unrealistically small sets, that we cannot afford to describe most sets by enumeration, and that we must be able to make general statements about classes of states. We then introduce an interpretation of classical predicate logic in which all expression denotations are finite sets and we show them how to use predicate calculus to characterise sets, including functions and relations. The logic that we use allows partial functions (defining all primitive predicates on undefined values to be <u>false</u>). It is important to provide numerous examples in which the students use predicate logic to characterise the states of something real. Arrays (viewed as partial functions) provide a rich source of examples such as, "Write a predicate that is <u>true</u> if array A contains a palindrome of length 3". Again, it is important to show the use of the mathematics to say important things about programs, and to teach them to use, as contrasted to prove theorems about, logic. The interpretation of logic that we use is described in [1].

4.4   Programs as "initial states"

We provide a brief, and unconventional, treatment of programming as picking the initial state of a finite state machine. This is necessary when one wants to explain such concepts as table driven programs, interpreters, etc. At this point, we point out von Neumann's chief insight in the area of computer design was the interchangeability of program and data.

4.5   Programs as descriptions of state-sequences

We then give a more conventional view of programs as descriptions of a sequence of state changes. Each program, given an initial state, describes one or more sequences of state changes. This concept is presented abstractly, we do not give any programming language notation for describing the sequences.

4.6   Programs as functions from starting-state to stopping-state

After pointing out that programs can be characterised as either *terminating* or *non-terminating* we indicate that this first course focuses on programs that are intended to terminate after computing some useful values. We then show that the most important characteristics of programs can be described by a mathematical relation between the starting-states and the stopping-states. The exact model used, LD-relations, is described in [2] or [3]. Here too, it is essential to provide examples in which the students use relations to describe distinct sets of sequences that are equivalent in the sense of having the same set of (start-state, final-state) pairs.

4.7   Tabular descriptions of functions and relations

We extend the notation of predicate calculus by introducing 2-dimensional tableaux, which we call tables, whose entries are predicate expressions or terms. We show that these are equivalent to more conventional notation, but easier to read. Students are given many examples in which we describe mathematical functions using these tables [7].

4.8   Teaching programming with this mathematical background

The remainder of the course is devoted to teaching students to program. All programs are introduced, not with a natural language description, but with a mathematical description of the required behaviour. The simple programming notation that is used (essentially that in [3]) is defined using the mathematical concepts above. We begin with very simple programs and continue, always using the same discipline, to cover more complex engineering problems. Homework assignments are specified using the tabular notation. Students are shown how to systematically determine if a program in this notation covers all cases and does the right thing in each case. Although, we never talk of "correctness proofs" we do use correctness concepts to explain a program. For example, we usually identify an "invariant" when explaining a loop, and use a monotonically decreasing quantity to convince students that a program will terminate.

5 The mathematics needed for software engineering

For many years I have taught a course entitled "Software Engineering" usually to students in the third or fourth year of university. Although the course has a significant design and pragmatic content, it has also been necessary to teach some mathematical concepts. Generally, Computer Science students have inadequate mathematical preparation for the course; they have learned too much theoretical computer science and too little about fundamental mathematics. However, the preparation of my Computer Engineering students seems even worse. They have learned lots of mathematics, but not the most relevant mathematics. In this section, I will describe the mathematical basis of my software engineering class. The class covers the "standard" software engineering topics and students are asked to do practical exercises, but the basic message is that they must produce a sequence of documents whose contents must be representations of key mathematical functions. The approach is basically that in [4]. To get maximum benefit from the course, students should already be familiar with the concepts described in the previous section. Usually, they have not had the necessary exposure, and much of the course must be devoted to mathematics.

5.1 How can we document system requirements?

A critical step in documenting the requirements of a computer system is the identification of the environmental quantities to be measured or controlled and the representation of those quantities by mathematical variables. The environmental quantities include: physical properties (such as temperatures and pressures), the readings on user-visible displays, administrative information (such as the number of people assigned to a given task), and even the wishes of a human user. These must be denoted by mathematical variables in the way that is usual in engineering. That association must be carefully defined, and coordinate systems, signs etc. must be unambiguously stated.

It is useful to characterise each environmental quantity as either monitored, controlled, or both. *Monitored* quantities are those that the user wants the system to measure. *Controlled* quantities are those whose values the system is intended to control. If needed, time can be treated as a monitored quantity. In the sequel, we will use "m_1", "m_2", $\cdots$, "m_p" to denote the monitored quantities, and "c_1", "c_2", $\cdots$, "c_q" to denote the controlled ones. Because it is often the case that a system is intended to both monitor and control certain quantities, these lists might have variables in common.

Each of these environmental quantities has a value that can be recorded as a function of time. When we denote a given environmental quantity by "v", we will denote the time-function describing its value by "v^t". Note that v^t is a mathematical function whose domain consists of real numbers; its value at time t is denoted by "$v^t(t)$".

The vector of time-function $(m_1^t, m_2^t, \cdots, m_p^t)$ containing one element for each of the monitored quantities, will be denoted by "$\underset{\sim}{m}^t$"; similarly $(c_1^t, c_2^t, \cdots c_q^t)$ will be denoted by "$\underset{\sim}{c}^t$".

5.1.1 Relation NAT

The environment, i.e. nature and previously installed systems, place constraints on the values of environmental quantities. These restrictions may be documented by means of a relation, which we call NAT. It is defined as follows:

- domain (NAT) is a set of vectors of time-functions containing only the instances of $\underset{\sim}{m}^t$ allowed by the environmental constraints,

- range (NAT) is a set of vectors of time-functions containing only the instances of $\underset{\sim}{c}^t$ allowed by the environmental constraints,

- $(\underset{\sim}{m}^t, \underset{\sim}{c}^t) \in$ NAT if and only if the environmental constraints allow the controlled quantities to take on the values described by $\underset{\sim}{c}^t$, when the values of the monitored quantities are described by $\underset{\sim}{m}^t$.

NAT is not always a function; if NAT is a function the computer system will not be able to vary the values of the controlled quantities without effecting changes in the monitored quantities.

5.1.2 Relation REQ

The computer system is intended to impose further constraints on the environmental quantities. The permitted values may be documented by means of a relation, which we call REQ. It is defined as follows:

- domain (REQ) is a set of vectors of time-functions containing the instances of $\underset{\sim}{m}^t$ allowed by environmental constraints,

- range (REQ) is a set of vectors of time-functions containing only those instances of $\underset{\sim}{c}^t$ considered permissible,

- $(\underset{\sim}{m}^t, \underset{\sim}{c}^t) \in$ REQ if and only if the computer system may permit the controlled quantities to take on the values described by $\underset{\sim}{c}^t$, when the values of the monitored quantities are described by $\underset{\sim}{m}^t$.

REQ is usually not a function because the application can tolerate "small" errors in the values of controlled quantities.

5.1.3 Requirements feasibility

Because the requirements should specify behaviour for all cases that can arise, it should be true that,

$$\text{domain (NAT)} \subseteq \text{domain(REQ)}. \tag{5.1}$$

The relation REQ can be considered *feasible with respect to NAT* if (5.1) holds and

$$\text{domain (REQ} \cap \text{NAT)} = (\text{domain (REQ) domain (NAT))}. \tag{5.2}$$

Feasibility, in the above sense, means that nature (as described by NAT) allows the required behaviour (as described by REQ); it does not mean that the functions involved are computable or that an implementation is practical.

Note that (5.1) and (5.2) can be reduced to:

$$\text{domain (REQ} \cap \text{NAT)} = \text{domain (NAT)}. \tag{5.3}$$

5.2 How can we document system design?

During the system design two additional sets of variables are introduced: one represents the *inputs*, quantities that can be read by the computers in the system; the other represents the *outputs*, quantities whose values are set by the computers in question. These variables are associated with input and output registers on the computers in the system; their values will also be described by time-functions.

In the sequel we assume that $\underset{\sim}{m}^t$ and $\underset{\sim}{c}^t$ are defined as in Section 4.2.

5.2.1 Relation IN

Let "$\underset{\sim}{i}$" denote the vector $(i_1^t, i_2^t, \cdots, i_r^t)$ containing one element for each of the input registers. The physical interpretation of the inputs can be specified by a relation IN, defined as follows:

- domain (IN) is a set of vectors of time-functions containing the possible instances of $\underset{\sim}{m}^t$,

- range (IN) is a set of vectors of time-functions containing the possible instances of $\underset{\sim}{i}^t$,

- $(\underset{\sim}{m}^t, \underset{\sim}{i}^t) \in$ IN if and only if $\underset{\sim}{i}^t$ describes possible values of the inputs when $\underset{\sim}{m}^t$ describes the values of the monitored quantities.

IN describes the behaviour of the input devices. It is a relation rather than a function because of imprecision in the measurements. It must be the case that,

$$\text{domain (NAT)} \subseteq \text{domain (IN)},$$

because the input device must transmit some value for every condition that can occur in nature.

5.2.2 Relation OUT

Let "$\underset{\sim}{o}^t$" denote the vector $(o_1^t, o_2^t, \cdots, o_s^t)$ containing one element for each of the output registers. The effects of the outputs can be specified by a relation OUT, defined as follows:

- domain (OUT) is a set of vectors of time-functions containing the possible instances of $\underset{\sim}{o}^t$,

- range (OUT) is a set of vectors of time-functions containing the possible instances of $\underset{\sim}{c}^t$,

- $(\underset{\sim}{o}^t, \underset{\sim}{c}^t) \in$ OUT if and only if $\underset{\sim}{c}^t$ describes possible values of the controlled quantities when $\underset{\sim}{o}^t$ describes the values of the output quantities.

OUT describes the behaviour of the output devices. It is a relation rather than a function because of device imperfections.

5.3 How can we document software requirements?

The software requirements are determined by the system design document and the system requirements document. The *software requirements document* can be seen as a combination of those two documents. It would contain the relations NAT, REQ, IN, and OUT.

In the sequel we assume that REQ is feasible with respect to NAT, and that $\underset{\sim}{m}^t$, $\underset{\sim}{c}^t$, $\underset{\sim}{i}^t$, and $\underset{\sim}{o}^t$ are defined as in previous sections.

5.4 Relation SOF

The software will provide a system with input-output behaviour that can be described by a relation, which we call SOF. It is defined as follows:

- domain (SOF) is a set of vectors of time-functions containing the possible instances of $\underset{\sim}{i}^t$,

- range (SOF) is a set of vectors of time-functions containing the possible instances of ϱ^t,

- $(\underset{\sim}{i}^t, \underset{\sim}{\varrho}^t) \in$ SOF if and only if the software could produce values described by ϱ^t when the inputs are described by $\underset{\sim}{i}^t$.

SOF will be a function if the software is deterministic.

5.4.1 Software acceptability

For the software to be acceptable, SOF must satisfy[1]:

$$\forall \underset{\sim}{m}^t \forall \underset{\sim}{i}^t \forall \varrho^t \forall \underset{\sim}{c}^t [\text{IN}(\underset{\sim}{m}^t, \underset{\sim}{i}^t) \wedge \text{SOF}(\underset{\sim}{i}^t, \varrho^t) \wedge \text{OUT}(\varrho^t, \underset{\sim}{c}^t) \wedge \text{NAT}(\underset{\sim}{m}^t, \underset{\sim}{c}^t) \Rightarrow \text{REQ}(\underset{\sim}{m}^t, \underset{\sim}{c}^t)] \tag{5.4}$$

Note, that if one or more of the predicates $\text{IN}(\underset{\sim}{m}^t, \underset{\sim}{i}^t)$, $\text{OUT}(\varrho^t, \underset{\sim}{c}^t)$, or $\text{NAT}(\underset{\sim}{m}^t, \underset{\sim}{c}^t)$ are false, then any software behaviour will be considered acceptable. For example, if a given value of $\underset{\sim}{m}^t$ is not in the domain of IN, the behaviour of acceptable software in that case is not constrained by (5.4).

If we assume that relations REQ, IN, OUT, and SOF are functions, we can use functional notation to rewrite (5.4) as follows:

$$\forall \underset{\sim}{m}^t [\underset{\sim}{m}^t \in \text{domain (NAT)} \Rightarrow \text{REQ}(\underset{\sim}{m}^t) = \text{OUT}(\text{SOF}(\text{IN}(\underset{\sim}{m}^t)))] \tag{5.5}$$

The writers of the requirements document must describe the relations NAT, REQ, IN, OUT. The implementors determine SOF and verify (5.4) or (5.5). A document of this type may require natural language in the description of the environmental quantities, but can otherwise be precise and mathematical. The use of natural language in the definition of the physical interpretation of mathematical variables is unavoidable and quite usual in engineering.

5.5 How can we document software behaviour?

Although the software requirements document fully represents the requirements that the software must meet, it may allow observable differences in behaviour. It will often be desirable to specify a subset of the behaviours allowed by the requirements document for actual implementation. In this way designers will make certain decisions that might otherwise have been

---

[1] In the following the universes from which $\underset{\sim}{m}^t$, $\underset{\sim}{c}^t$, $\underset{\sim}{i}^t$ and ϱ^t are drawn are assumed to include all vectors of time-functions

left for the programmers. The relation SOF can be described in a separate document known as the *software behaviour specification*. This document is especially important for multiple-computer systems because it will define the allocation of tasks to the individual computers in the system. For computer networks, or multi-processor architectures, one may see a hierarchy of software behaviour specifications with upper level documents assigning duties to groups of computers, and the lowest level documents, detailing the responsibilities of individuals computers.

5.6 How can we document black-box module interfaces?

Most modern computer systems require software of such size and complexity that it cannot be completed by a single person in a few weeks. For many reasons it is desirable to decompose the software construction task into a set of smaller programming assignments. Each assignment is to produce a group of programs (cf. Section 4.8) which we call a *module*. We view each module as implementing one or more finite state machines, frequently called *objects* or *variables*. A description of the module interface is a black-box description of these objects.

Writing software module interface specifications is similar to documenting software requirements but some simplifications are possible. Many software modules are entirely internal; there are no environmental quantities to monitor or control and all communication can be by means of external invocation of the module's programs. Moreover, the state set of a software module is finite, and state transitions can be treated as discrete events. For most such modules, real-time can be neglected because only the sequence of events matters. This allows us to replace the general concept of time-function by a sequence describing the history in terms of discrete events; we call these sequences *traces*.

We identify a finite subset of the set of possible traces, which we call *canonical traces*. Every trace is equivalent[2] to a single canonical trace. *Trace assertion specifications* comprise three groups of relations:

1. Functions whose domain is a set of pairs (canonical trace, event) and whose range is a set of canonical traces. The pair $((T_1, e), T_2)$ is in the function if and only if the canonical trace T_2 is equivalent to the canonical trace T_1 extended with "e". These functions are known as *trace extension functions*[3].

2. Relations whose domain contains all the canonical traces and associate each canonical trace with a set of values for the output (externally visible) variables.

---

[2] Two traces are *equivalent* if they have the same effect on future behaviour of the object.

[3] A trace extension function is sometimes called a *reduction* function.

3. Functions whose domain is the set of values of the output variables and whose values define the information returned by the module to the user of the module.

5.7 How can we document internal module design?

Each module has a private data structure and one or more programs. We propose to document the design sufficiently precisely that its correctness can be verified. The internal documentation of a module contains three types of information:

1. A complete description of the data structure, which may include objects implemented by other modules.

2. A function, known as the *abstraction function*, whose domain is a set of pairs (object name, data state), and whose range is a set of canonical traces for objects created by the module. The pair $((on, ds), T)$ is in this function if and only if a trace equivalent to T describes a sequence of events affecting the object named on that could have resulted in it being in data state ds.

3. An LD-relation [2,3], often referred to as the *program function*, specifying the behaviour of each of the module's programs in terms of mappings from data states before the program execution to data states after the execution.

6 Do we need new mathematics or merely new representations?

There is something in the above that will be disturbing, perhaps even annoying, to many people. We have managed to make precise mathematical statements about software engineering using <u>classical</u> mathematical concepts. We have not used <u>any</u> of the relatively new "specification languages", which have been developed especially for software engineering applications. We have even been able to talk about the real-time characteristics of systems without introducing any changes in our logic for that purpose; we have dealt with real-time using the traditional engineering approach, the use of functions whose range and domain are taken from the set of time-functions. I have studied the new "formal methods" and simply do not see how they add value. It seems to me that the mathematics needed by engineers to understand software is very close to the classical mathematics that was developed before Computer Science became an identified "discipline". In [5] I presented some serious doubts about the direction taken by Computer Science; this paper presents further grounds for those doubts.

On the other hand, when we tackle real software engineering problems, such as the A-7 Onboard Flight Program [8], or the Darlington Nuclear Plant [9], we find a need, not for new basic concepts but for new notations. The use of conventional, one-dimensional, notation to describe functions and relations resulted in pages of repetitive formulae that were hard to parse. It is for this reason that we have introduced the multidimensional notations, first used in [8] and described in [7]. The new specification languages have not deviated in a significant way from the one-dimensional notation that is traditional in mathematics. Our experience suggests that the semantic issues are not the serious ones. On the other hand, new, multi-dimensional notation, with classical semantics, has proven very practical.

Acknowledgements

These thoughts have been strongly influenced by H.D. Mills and N.G. de Bruijn. Much of the text was taken from a paper written jointly with Professor Jan Madey of Warsaw University ([4]).

Bibliography

1. Parnas, D.L. (1992). Predicate logic for software engineering. *CRL Report 241*, McMaster University, Telecommunications Research Institute of Ontario (TRIO), February, Accepted by IEEE Transactions on Software Engineering.

2. Parnas, D.L. (1983). A generalized control structure and its formal definition". *Communications of the ACM, 26, 8*, August, 572-581.

3. Parnas, D.L. and Wadge, W.W. (1986). Less Restrictive constructs for structured programs. *Technical Report 86-186*, Queen's, C&IS, Kingston, Ontario, Canada, October.

4. Parnas, D.L. and Madey, J. (1991). Functional documentation for computer systems engineering (Version 2). *CRL Report 237*, McMaster University, TRIO, September, 14 pages.

5. Parnas, D.L. (1990). Education for computing professionals". *Proceedings of International Conference on Computing and Information*, ICCI'90, Niagara Falls, Ontario, 23rd-26th May. Published in Advances in Computing and Information, Editors: S.G. Akl, F. Fiala and W. Koczkodaj, Canadian Scholars' Press Inc., 11 pages, (ISBN 0-921627-70-X).

6. Nederpelt, R.P. (1987). De taal van de wiskunde. Versluys Uitgeverij bv - Almere, NL.

7. Parnas, D.L. (1992). Tabular representation of relations. *CRL Report 260*, McMaster University, TRIO, October.

8. Heninger, K.L., Kallander, J., Parnas, D.L. and Shore, J.E. (1978). Software requirements for the A-7E aircraft. *NRL Memorandum Report 3876*, United States Naval Research Laboratory, Washington D.C., November, pp. 523.

9. Parnas, D.L., Asmis, G.J.K. and Madey J. (1991). Assessment of safety-critical software in nuclear power plants. *Nuclear Safety,32, 2*, pp. 189-198.

Formal demonstration of equivalence of source code and PROM contents: an industrial example

D.J. Pavey and L.A. Winsborrow

Nuclear Systems Branch, Nuclear Electric, Gloucester

Abstract High level language compilers and linkers are vital to the modern production process for safety-critical software. A technique for demonstration of the correctness of the translation process for a specified set of source programs is outlined, and illustrated by an example of formal verification of a selected language feature ('pointer based variables'). A non-compliance between source and PROM code, revealed by the technique, is discussed.

1   Introduction

Not many years ago, the standard method of producing a safety-critical system was to choose a processor and then program directly in assembly language which mapped directly to the processor's instruction set.

Users now demand much greater sophistication from safety-critical systems in terms of facilities offered and flexibility of response; the extra functionality is achieved by incorporating ever larger and more complex algorithms into the software. Such systems have been made possible by the development of high-level languages which increase productivity and incorporate strong typing and other restrictions, helping to assure the integrity of the final product.

These advances have inevitably necessitated greater dependence on complicated development tools, principally compilers and linkers. Such tools are a vital link in the production chain; if they contain errors, incorrect code may be generated. If the input space of the dynamic tests fails to cover the erroneous cases, the errors may remain dormant in the system for some time.

Some practitioners argue that the reliability of a tool may be judged from the extent of its previous use "in the field"; the strength of this argument is somewhat reduced by the observation that an established tool has usually passed through a large number of versions and been subject to many modifications and it is known that error correction sometimes introduces other errors.

225

An alternative approach is the formal verification of the tool (the compiler in this case). Currently this has only been undertaken as an academic exercise, rather than upon a compiler in widespread commercial use [1,2].

A third approach is to demonstrate the correctness of the translation for the compilation process performed on the programs incorporated in a given safety-critical system, i.e. for a specified subset of the set of all possible systems; this approach is the subject of the current paper. The background to the project is the assessment of the Primary Protection System (PPS) for the Sizewell B Nuclear Power Station [3]. The implementation of the necessary utilities has been described more fully elsewhere [4]; this paper outlines the approach and then attempts to convey a deeper understanding of the formal processes by concentrating on an example of formal verification of compiled code as illustrated by a selected language feature. Finally a non-compliance between source and PROM code is discussed.

2 Approach

Comparison of high-level language with machine code is a non-trivial problem for several reasons. Firstly, the mapping between instructions is usually one-to-many. Secondly an optimising compiler rearranges or deletes blocks of code, often taking an entire procedure as the range for the optimisation. Thirdly some source language features are not explicitly represented in the object code (e.g. blocks) or cannot be unambiguously identified (e.g. FOR loops). For these reasons reconstruction of the source followed by direct comparison was not feasible. An alternative possibility was the symbolic evaluation of all outputs in terms of inputs for corresponding sections of both source and object code; the resulting algebraic expressions could then be compared and any discrepancies noted. This approach was adopted, the process being automated by use of the MALPAS static analysis tool [5].

Specifically, the goal was to perform source/code comparison on PROM files targeted on the Intel i8086 processor family. The process is outlined in Figure 1.

The object code was reconstituted from PROM code to a form of ASM-86 using a series of specially-written utilities. The source was scanned to determine additional information such as the order of declaration and types of variables together with initialisation values. This information on the program objects was stored in a 'name table' for reference during the remainder of the process. Both source and reconstituted ASM-86 were translated into the MALPAS Intermediate Language (IL). Translators are commercially available for both ASM-86 [6] and the source language, PL/M-86 [7]. In principle, the translators can be viewed as defining a denotational semantics for the two languages in terms of an extended version of the MALPAS

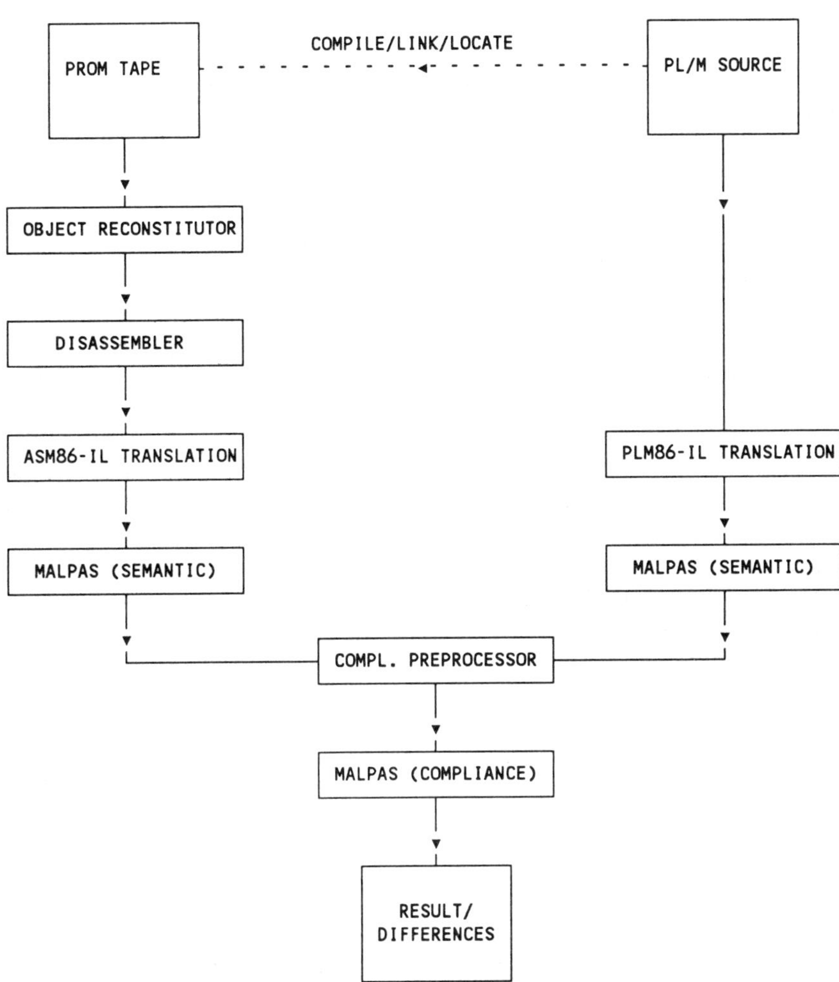

Figure 1. Outline of the Source/PROM comparison process

IL 'algebra'; although in practice a high level of formality, for both the semantic mapping and the extended 'algebra' itself, is yet to be established.

The MALPAS IL reader was then used to transform both versions of the program into a graph of nodes and arcs, together with their associated semantics. The MALPAS semantic analyser then derived symbolic expressions for the "outputs" of each arc in terms of the "inputs". Finally the MALPAS compliance analyser was used to perform a comparison between the symbolic expressions obtained from each arc of the transformed source and the equivalent arc of the transformed object code. Clearly this method depends on the source and object code having the same 'structure'; this limitation, among others, is discussed briefly in [4].

The use of the MALPAS tool-set conferred the advantage that large quantities of object code could be verified against source with little human intervention and a high degree of confidence in the results. Some language features required further transformation in order to achieve similarity of representation. These transformations could be performed without loss of rigour by extending the standard MALPAS algebraic simplifier with additional application specific rewrite rules. Such rules are open to inspection for rigour and correctness, and the results of applying them are predictable and reproducible. The rules are generally very simple and independent, but have not been formally proved, and in some cases have a more pragmatic than formal basis. The ability to create rewrite rules is an integral part of the MALPAS Intermediate Language, in which they are known as 'replacement' rules.

The next section of the paper describes the development and application of one such transformation.

3   Example

A feature of the PL/M-86 language is the use of 'pointer' variables to give access to data structures whose location in memory is not known until execution time. This can present difficulties for any formal verification because of the problem of aliasing but, as indirect addressing is explicit in the source language, this is not really an issue for the subject of this paper. We are only concerned with demonstrating that the semantics of the source language have been correctly preserved in the executable code.

This section shows how the comparison process deals with indirect data access by symbolic execution of the sequence of assembler instructions required, and by using application specific rewrite rules to simplify the resulting expression into a form that can be compared with the source language. A worked example is used to show some of the principal intermediate forms of the code representation. The example also shows how loops are handled. Other features such as the handling of floating point operations and procedure calls are illustrated in [4]. The example presented here is deliberately chosen to be very simple to illustrate a few basic principles. Very much larger and more complex examples were encountered in the PPS code analysed, but it would be impractical to present more than the 'tip of the iceberg' here.

3.1   The source and compiled code

The PL/M-86 source (A.1) consists of the declaration of a procedure to sum the elements of a vector of integer values. It is passed two parameters: a pointer to the vector (vector_ptr) and the index (vend) of the last value to be summed, and it returns the sum of the first 'vend+1' vector elements starting from item zero. We are not concerned here with whether this short PL/M-86 procedure correctly implements the 'requirement' outlined informally above. For the Sizewell B protection system software, that was a separate and very significant task which also made extensive use of the MALPAS tool-set [3]. Our present task is to accept the source as a correct verified statement of the requirement, and to demonstrate that it is correctly implemented by the executable code in the protection system PROMs.

This example is further simplified by only comparing the source with the object file generated by the compiler - the reversal of the linking and locating processes is not illustrated. The object file (A.2 - converted to hex-ascii format for readability) contains a series of records representing the binary coded machine instructions.

This is disassembled and, using information from the name table, most of the memory references are resolved to give a code representation (A.3) similar to that which appears in the PL/M-86 compiler listing (A.4). The only memory reference which has not been identified by name is the read from the based array 'vector' (location 001E). This cannot be resolved by the information in the single instruction at that address, since it depends on the contents of the registers ES, BX and SI, which are loaded in earlier instructions. In more realistic examples, these registers may be loaded many instructions earlier. Even in this example, it is necessary to look back six instructions in order to determine the contents of SI in terms of variable names. The resolution of this memory reference is deferred until after symbolic execution has combined all the required information in a single expression, as described below.

3.2   Modelling the compiled code

The translation (A.5) of the assembler code into MALPAS Intermediate Language (IL) is significantly longer, involving the use of several IL statements to model a single assembler statement. The features of the IL model are determined by the commercially available translator [6], supplemented by features provided specifically for this project. As an example, some explanation of the model of indirect memory access is given below.

Access to memory by address is represented using the 'monolithic memory model', that is, the whole of memory is treated as a single large array of integers, indexed by memory address. Arrays are modelled in MALPAS IL using the type extension 'array'. A function 'update' represents writing to an array:

```
FUNCTION update (integer-array, integer, integer)
                : integer-array;
```

by mapping the unmodified array, the array index, and the value to be written respectively onto the modified array. An operator '!' represents reading from an array:

```
INFIX ! (integer-array, integer) : integer;
```

by mapping the array and the array index onto the element value. Rewrite rules are used to 'tell' MALPAS the semantics of these operators as follows:

```
REPLACE (a : integer-array; i, j : integer;
         n : integer)
   update (a, i, n) ! j     BY n   IF i = j,
                            BY a!j IF i/= j;
```

which allows the recovery of an item earlier written to an array element, and the elimination of data written to an element in which we are not interested; and also:

```
REPLACE (a : integer-array; i, j : integer;
         m, n : integer)
   update (update (a, i, m), j, n)
 BY update (a, j, n)                     IF i = j,
 BY update (update (a, j, n), i, m) IF i>j;
```

which allows the elimination of data overwritten by a subsequent write to the same array element, and the ordering of data according to index number, which often facilitates the operation of the first clause of this rule.

For the purposes of indirect memory access then, the processor memory is declared as an array:

```
  mem : integer-array;
```

and in our example (A.5, label 00110:) the read from the based array is modelled as:

```
ax := mem ! (es*16 + bx + si + 0);
```

The array index here is an expression which makes explicit the addressing strategy for the Intel 8086 family. The 20-bit address of a memory location is constructed from a 16-bit segment register, in this case ES, the 'extra segment', which is normally used to access based variables, and a 16-bit offset, here the combination of the BX and SI registers. The segment register is left-shifted four places and added to the offset to create the required 20-bit memory address.

3.3 'Decompilation'

The IL model of the assembler code (A.5) is submitted to the MALPAS semantic analyser, which performs symbolic execution of each section of code, i.e. of each arc between marked nodes. This gives an algebraic expression for the value of each variable at the end of the arc in terms of the values of the variables at the start of the arc.

Note that in our example, a marked node has been inserted automatically at the loop head. This ensures that all program sections analysed are loop free, so avoiding the problems associated with formal verification of code containing loops. This strategy relies on the ability to identify the corresponding loop constructs in both the source and the low level code. This was fairly straightforward for PL/M-86, but sometimes needed manual intervention to assist the rather unsophisticated empirical algorithm used for automatic identification.

Symbolic execution of the arc comprising the body of the loop gives the following expression for 'sum':

```
sum : =    sum
        + mem ! (  2 * i + lowword(vector_ptr)
                + 16 * highword(vector_ptr));
```

The reordering of terms is due to the notion of normal form built into the MALPAS algebraic simplifier.

Further rewrite rules were constructed to simplify expressions like these, for example:

```
REPLACE (a,b : integer)
    b + lowword (a) + 16*highword (a)
  BY adr (a) + b;
```

where 'adr' is an abstract function which maps a 32-bit 'pointer' value onto its corresponding memory address.

Including this in the rule base for the semantic analysis of our example yields the following expression for 'sum':

```
sum : = sum + mem ! (2 * i + adr(vector_ptr));
```

We now have all the information needed to dereference this memory read in a single expression. All expressions involving 'mem' are identified and parsed to extract the elements of the address. The name table can then be used to find the name and type of the variable based on the pointer, and in this case the expression for 'sum' is replaced by the following:

```
sum : = sum + vector ! (i);
```

Note that the index 'i' is now no longer multiplied by 2. This is because the 8086 addresses each byte of memory separately, so 'mem' is a byte array, whereas 'vector' is an integer array, each integer occupying a pair of bytes.

The semantic analyser also derives expressions for the contents of each register, and each intermediate variable introduced by the translator, at the end of each program arc. These are eliminated since they are implementation details which have no counterpart in the source code. The expressions for each of the remaining program variables should have no references to registers or intermediate variables, unless the compiler has tracked the contents of registers across marked node boundaries. This is not a problem for PL/M-86 loop heads and procedure calls, where marked nodes are inserted automatically. Special care is required if additional marked nodes need to be inserted manually.

This results in a high-level representation (A.6) of the behaviour of the code in the PROM, in terms of the algebraic expression for each program variable written to in each arc of the program. This is the nearest we come to 'decompiling' the low level code. Each arc is represented for convenience of comparison as a separate IL procedure. We now need to derive a representation of the source code behaviour in the same format, so that we can then compare the code for corresponding arcs.

3.4   Comparison with the source

The IL model of the source code (A.7) is generated using the commercially available translator [7], again supplemented by features specific to this project such as the automatic insertion of marked nodes at loop heads and procedure calls. This is submitted to the MALPAS semantic analyser, and the results reformatted as before (A.8).

It now remains to perform the final comparison between the two representations of the program (A.6 and A.8). In this case it is trivial to see by eye that they are equivalent, but even here there are slight differences in format which make a direct character-by-character comparison impossible. For more realistic examples the differences can be much more significant. The MALPAS tool-set provides the facility to check a program for compliance with a formal specification in terms of pre- and post-conditions. In order to use this facility, the representation of the source code for each arc is converted to the formal specification format, and the representation of the PROM contents is appended to this to make a combined IL text (A.9).

The compliance analyser generates a 'threat' condition for each program arc as follows:

```
_threat : =      <PRE condition>
           AND   <arc predicate>
           AND NOT <POST expression>
```

where <arc predicate> asserts the status of the variables at the end of the arc. Since the precondition is null ('true') in this example, this effectively simplifies to:

```
_threat : =      <PROM code predicate>
           AND NOT <source code predicate>
```

which simplifies to 'false' if the PROM code implies the source code.

A final threat statement (A.10) is extracted from the results of the compliance analysis, as a record of a successful demonstration of correct implementation.

To illustrate the effect of a compiler error, the PROM version of the code was seeded with an error by removing the index register from the indirect memory read, i.e:

```
MOV AX,ES:(BX)+(SI)+00000
```

was changed to

```
MOV AX,ES:(BX)+00000
```

resulting in the following threat statement:

```
  vector_sum_112_112:
_threat := vend >= i AND vector ! 0 <> vector ! i
```

which clearly highlights the difference between the two versions.

4 Result of application

The method and tool-set described above were developed as one component of the assessment process of the Sizewell B Primary Protection System, which contains over 85000 lines of PL/M-86 code. Source/code comparison has been applied to a pre-operational version of the code and is being repeated as necessary to verify subsequent changes.

Application of the method revealed one code generation error by the compiler, as well as an invalid use of a pointer comparison which the compiler failed to detect. A few non-compliances with the project coding standard were also noted and reported as a result of the close examination of the code which took place in the course of the project.

The code generation error was a failure to keep track of register allocation (a classic compiler error) which effectively resulted in the removal of a comparison; instead of two unlike items being compared, a register was compared with itself, and naturally the equality was always true. The error arose in the comparison of a pointer (occupying two 16-bit words) with zero. The two halves of the pointer were placed in the DX and ES registers, and the SI and DI registers were set to zero for the comparison. The instruction set does not contain an instruction for direct comparison between the ES register and another register, so the contents of the ES register were moved to DI, the compiler apparently forgetting that DI had already been allocated to hold zero. Hence DI is compared with itself, and always gives equality. Consequently the results from semantic analysis of the ASM-86-derived code were not as expected, and therefore an intractable threat was obtained when these results were compared with the semantic analysis output of the PL/M-86 derived code (see Appendix B).

The error was present only in one module, but occurred eleven times within that module, always in the comparison of pointers. The developers (Westinghouse, Pittsburgh) were able to show that, given the known positions of certain program components within memory, the error would not lead to a fault in the Sizewell B PPS. The erroneous comparison would only lead to an incorrect result where the upper half (selector) of the pointer is non-zero and the lower half (offset) is zero; given the pointer values under examination such a situation could not arise. For further assurance, the developers decided to implement a change in the operational version of the software in order to circumvent the construct which led to the error. This has resulted in appropriate reverification and testing.

It is of interest that the error was manifested in code compiled using version 3.3 of the compiler; it could not be reproduced using version 3.1. This suggests strongly that the existence of a large customer base for a development tool is insufficient evidence for the tool's correctness; the Intel PL/M-86 compiler is widely used but this was no defence against the introduction of an error during maintenance of the tool.

Another shortcoming in the compiler was revealed by the inclusion of a comparison between pointers using the 'less than' ($<$) operator. The Intel PL/M-86 Programmers Guide forbids the use of any comparison between pointers other than equality ($=$) or inequality ($<>$). Unfortunately the compiler gave no warning when an illegal construct occurred; instead it generated code for an unsigned double-word comparison. Such a comparison applied to Intel 8086 pointers could well give the opposite result from that intended, because of the special nature of pointer representation. Source/code comparison generated an unresolvable threat which drew attention to the problem. Again the developers were able to show that the actual pointer values involved would not lead to errors.

The results of this comparison exercise have implications for the reliance which can be placed on automatic transformation tools, such as compilers and linkers, in systems critical for safety. This issue is the subject of some debate, in the light of the strong requirements of important standards such as the Interim DEF STAN 00-55 [8]. Both the faults are of a class which could remain undetected by full testing; a change later in the system's lifetime giving rise to changes of component addresses, implemented after limited testing, could lead to system failure.

5 Conclusion

This paper has described a method of demonstrating correctness of the source to PROM code translation process, correctness being demonstrated for a specified source program as one component of a specified system rather than for all possible source programs in any linkage combination. The process has been automated and applied to assessment of a real non-trivial system and one compiler code-generation error was detected together with one area in which the compiler did not enforce a restriction of the language. Neither of these deficiencies compromise the present configuration. Confidence in the system was enhanced by the inclusion of source/code comparison in the assessment programme; it is seen by the licensor as an important part of the safety case [9].

Acknowledgements

The authors acknowledge the contributions of A.R. Lawrence for the original proposal, and of I.M. Andrews in the object code reconstitution phase of the project.

The MALPAS tool and the MALPAS IL translators are supported and marketed by TA Consultancy Services Ltd, Farnham.

This paper is published with the permission of Nuclear Electric plc.

Bibliography

1. Stepney, S., Whitley, D., Cooper, D. and Grant, C. (1991). A demonstrably correct compiler. *BCS Formal Aspects of Computing*, *3*, 58-101.

2. Bowen, J. (1991). From programs to object code and back again using logic programming. *ESPRIT Redo project report 2487-TN-PRG-1044 version 0.3*, January.

3. Ward, N.J. (1993). The rigorous retrospective static analysis of Sizewell 'B' primary protection system software. *SAFECOMP'93: Proceedings of the 12th International Conference on Computer Safety, Reliability and Security*, Poland, Springer-Verlag, pp.171-181.

4. Pavey, D.J. and Winsborrow, L.A. (1993). Demonstrating equivalence of source code and PROM contents. *The Computer Journal*, *36*, 7, 654-667.

5. Rex, Thompson and Partners Limited (TA Consultancy Services Limited) (1991). *MALPAS User Guide, release 5.1, RTP/9039/01*, April.

6. Rex, Thompson and Partners Limited (TA Consultancy Services Limited) (1991). *ASM-86 to IL Translator User Guide, release 2.1, RTP/1019/02*, September.

7. Rex, Thompson and Partners Limited (TA Consultancy Services Limited) (1991). *PL/M-86 to IL Translator User Guide, release 2.1, RTP/9039/14*, January.

8. UK MoD. (1991). The Procurement of Safety Critical Software in Defence Equipment. *INT DEF STAN 00-55 (PART 1)/1*, April.

9. Hunns, D.M. and Wainwright, N. (1991). Software-based protection for Sizewell B: the regulator's perspective. *Nuclear Engineering International*, *36*, 38-40.

Appendix A: Example including access to a pointer-based variable

A.1. PL/M-86 source

```
$TITLE ('Based Example')

$ MEDIUM ROM OPTIMIZE(3) CODE XREF DEBUG LIST NOINTVECTOR
IMA_EXAMPLE:

DO;
```

```
VECTOR_SUM:
    PROCEDURE (vector_ptr, vend)                integer ;

    declare vector_ptr                          pointer ;
    declare vector based vector_ptr (*)         integer ;
    declare vend                                integer ;
    declare i                                   integer ;
    declare sum                                 integer ;

    sum = 0;
    DO i = 0 to vend;
        sum = sum + vector (i);
    END;

    return ( sum ) ;

END VECTOR_SUM;

END IMA_EXAMPLE;
```

A.2. Extract from object file in hex ascii format

```
LEDATA
0100 00
558B ECC7 0602 0000 00C7 0600 0000 00A1 0000 3946 047C
14D1 E0C4 5E06 8BF0 268B 0001 0602 00FF 0600 00EB E4A1
0200 5DC2 0600
```

A.3. Disassembled file

```
segment index=  01     start offset= 0000
0000            PUSH BP; start vector_sum
0001            MOV BP,SP
0003            MOV sum,#0000; word op
0009            MOV i,#0000; word op
000F @00002:    MOV AX,i
0012            CMP vend,AX
0015            JL @00001; short jump
0017            SAL AX,#1; word op
0019            LES BX,vector_ptr
001C            MOV SI,AX
001E            MOV AX,ES:(BX)+(SI)+00000
0021            ADD sum,AX
```

```
0025               INC i; word op
0029               JMP @00002; short jump
002B @00001:       MOV AX,sum
002E               POP BP
002F               RET #0006; intra end vector_sum
```

A.4. Assembly listing of object code produced by the compiler

```
                                                   ; STATEMENT # 2
                         VECTOR_SUM        PROC NEAR
        0000  55                   PUSH    BP
        0001  8BEC                 MOV     BP,SP
                                                   ; STATEMENT # 8
        0003  C70602000000         MOV     SUM,OH
                                                   ; STATEMENT # 9
        0009  C70600000000         MOV     I,OH
                 @1:
        000F  A10000               MOV     AX,I
        0012  394604               CMP     [BP].VEND,AX
        0015  7C14                 JL      @2
                                                   ; STATEMENT # 10
        0017  D1EO                 SAL     AX,1
        0019  C45E06               LES     BX,[BP].VECTOR_PTR
        001C  8BFO                 MOV     SI,AX
        001E  268B00               MOV     AX,ES:[BX].VECTOR[SI]
        0021  01060200             ADD     SUM,AX
                                                   ; STATEMENT # 11
        0025  FF060000             INC     I
        0029  EBE4                 JMP     @1
                 @2:
                                                   ; STATEMENT # 12
        002B  A10200               MOV     AX,SUM
        002E  5D                   POP     BP
        002F  C20600               RET     6H
                                                   ; STATEMENT # 13
                         VECTOR_SUM        ENDP
```

A.5. IL model of ASM-86 ready for semantic analysis

```
TITLE imaex;

[*** Rex, Thompson and Partners Ltd ASM86-IL
Translator Release 2.1 ***]
```

```
_INCLUDE/NOLIST "asmprelude21"
_INCLUDE/NOLIST "compr:rules.ila"

CONST stoff_vector_sum : integer = 0;
FUNCTION vector_sum_call
        (integer, integer) : external_world;
PROCSPEC vector_sum( OUT pcall : external_world
            INOUT fsp : integer  IN fstack : real-array
            IN stack_error, fcf, fzf : boolean
            IN ax, al, ah, bx, bl, bh, cx, cl, ch,
            dx, dl, dh : integer
                  INOUT sp : integer
            IN si, ds, di, bp, cs, es, ss, ip : integer
                  IN cf, of, sf, zf : boolean
                  IN mem, memds, port, stack : integer-array )
DERIVES sp  AS sp + 8,
        fsp  AS fsp - 0,
        pcall AS vector_sum_call
                (pointer_value (65536*stack!(sp+6)+stack!(sp+4)),
                stack!(sp+2));

PROC vector_sum;

VAR interrupt_status : interrupt_state;
VAR vector_sum_result : integer;

[ ASM-86 basic type variables ]
VAR i, sum, vector, vector_ptr, vend, temp, temp2 : integer;

    00010:      sp := sp - 2;                    [ PUSH BP ]
                stack := update(stack, sp, bp);

    00020:      bp := sp;                        [ MOV BP,SP ]

    00030:      sum := 0;                        [ MOV SUM,0 ]

    00040:      i := 0;                          [ MOV I,0 ]

      stack_error := bp /= sp + stoff_vector_sum;
112: [marked node (loop head)]
      bp := sp + stoff_vector_sum;
                ax := i;                  [@00002:   MOV AX,I ]
                al := low(ax);
                ah := high(ax);

    00060:      temp := vend - ax;               [ CMP VEND,AX ]
                cf := posword(vend) < posword(ax);
```

```
                    of := (temp > 32767) OR (temp < -32768);
                    sf := (temp < 0) AND (temp > -32769)
                    OR (temp > 32767);
                    zf := temp = 0;

  00070:            IF (sf XOR of) THEN                    [ JL @00001 ]
                       GOTO 150 [@00001]
                    ENDIF;

  00080:            ax := ax * 2;                          [ SAL AX,1 ]
                    al := low(ax);
                    ah := high(ax);

  00090:            bx := lowword(vector_ptr);   [LES BX,VECTOR_PTR ]
                    bl := low(bx);
                    bh := high(bx);
                    es := highword(vector_ptr);

  00100:            si := ax;                              [ MOV SI,AX ]

                                              [MOV AX,ES:[BX + SI +0] ]
  00110:            ax := mem ! (es*16 + bx + si +0);
                    al := low(ax);
                    ah := high(ax);

  00120:            temp := sum;
                    sum := sum + ax;                       [ ADD SUM,AX ]
                    cf := posword(temp) + posword(ax) > 65535;
  00130:            i := i + 1;                            [ INC I ]
                    of := i > 32767;
                    sf := (i < 0) OR (i > 32767);
                    zf := i = 0;

  00140:            GOTO 112 [@00002];                     [ JMP @00002 ]

  00150:            ax := sum;                 [ @00001:  MOV AX,SUM]
                    al := low(ax);
                    ah := high(ax);

  00160:            bp := stack ! sp;                      [ POP BP ]
                    sp := sp + 2;

  00170:            sp := sp + 2 + 6;                      [ RET 6 ]

            vector_sum_result := ax;
ENDPROC;
FINISH
```

A.6. Results of semantic analysis of the PROM code

```
PROC vector_sum_start_112;
IF true
THEN MAP
   i := 0;
   sum := 0;
ENDMAP ENDIF;
ENDPROC;


PROC vector_sum_112_112;
IF i <= vend
THEN MAP
   i := i + 1;
   sum := sum + vector ! (i);
ENDMAP ENDIF;
ENDPROC;


PROC vector_sum_112_end;
IF i > vend
THEN MAP
   vector_sum_result := sum;
ENDMAP ENDIF;
ENDPROC;
```

A.7. IL model of PL/M-86 ready for semantic analysis

```
[   IL Produced by PL/M-86 Translator Version 2.16b ]

TITLE ima_example;
_INCLUDE/NOLIST "plm$prelude"
_INCLUDE/NOLIST "il_rules"


FUNCTION vector_sum_call
        (pointer, integer) : external_world;
PROCSPEC vector_sum([Formal Parameter Section]
                   IN vector_ptr : pointer
                   IN vend : integer
  [Typed Procedure Result Section]
                   IN vector_sum_result : integer)
   IMPLICIT (OUT pcall : external_world)
DERIVES pcall AS vector_sum_call
               (vector_ptr, vend);
```

```
PROC vector_sum;
   VAR vector [BASED vector_ptr] : integer-array;
   VAR i : integer;
   VAR sum : integer;
   VAR limit_value, step_value : integer;
   sum := 0;
   i := 0;
112: [marked node (loop head)]
   LOOP
      limit_value := vend;
      step_value := 1;
      EXIT WHEN ((step_value <  0) AND (i < limit_value))
            OR  ((step_value >= 0) AND (i > limit_value));
      sum := sum + vector!(i);
      i := i + step_value
   ENDLOOP;
   vector_sum_result := (sum);
   STOP
ENDPROC
FINISH
```

A.8. Results of semantic analysis of the source code

```
PROC vector_sum_start_112;
IF true
THEN MAP
   i := 0;
   sum := 0;
ENDMAP ENDIF
ENDPROC;


PROC vector_sum_112_112;
IF vend >= i
THEN MAP
   i := i + 1;
   sum := sum + vector ! i;
ENDMAP ENDIF
ENDPROC;


PROC vector_sum_112_end;
IF vend < i
THEN MAP
   vector_sum_result := sum;
ENDMAP ENDIF
ENDPROC;
```

A.9. Combined file ready for final compliance analysis

```
TITLE ima_example;
_INCLUDE/NOLIST "plm$prelude"
_INCLUDE/NOLIST "il_rules"

FUNCTION vector_sum_call
          (pointer, integer) : external_world;



PROCSPEC vector_sum_start_112(
INOUT dummy_var : external_world
 OUT i : integer
 OUT sum : integer)
PRE (true)
POST ((true)
 --> ((i = (0)) AND (sum = (0))));

PROCSPEC vector_sum_112_112(
INOUT dummy_var : external_world
 IN vend : integer
 INOUT i : integer
 INOUT sum : integer
 IN vector [BASED vector_ptr] : integer-array)
PRE (true)
POST (('vend >= 'i)
 --> ((i = ('i + 1)) AND (sum = ('sum + 'vector ! 'i))));

PROCSPEC vector_sum_112_end(
INOUT dummy_var : external_world
 IN vend : integer
 IN i : integer
 OUT vector_sum_result : integer
 IN sum : integer)
PRE (true)
POST (('vend < 'i)
 --> ((vector_sum_result = ('sum))));




PROC vector_sum_start_112;
IF true
THEN MAP
```

```
     i := 0;
     sum := 0;
ENDMAP ENDIF;
ENDPROC;


PROC vector_sum_112_112;
IF i <= vend
THEN MAP
     i := i + 1;
     sum := sum + vector ! (i);
ENDMAP ENDIF;
ENDPROC;


PROC vector_sum_112_end;
IF i > vend
THEN MAP
     vector_sum_result := sum;
ENDMAP ENDIF;
ENDPROC;
FINISH
```

A.10. Final compliance statement

```
MALPAS Release 6.0.2 (2Mb)
18-AUG-1993 15:22 - Compliance Analysis

 Non-false threats from source-code comparison of: ima_example

All threats are 'false'.
```

Appendix B: Fault investigation process for register allocation error

This appendix outlines the investigation which led to the discovery of the register allocation error. It is not possible to reproduce the original files for reasons of copyright and available space.

B.1 Extract from PL/M-86 code

The line marked **** is wrongly compiled under certain conditions.

```
    declare (final_ptr,init_ptr)        pointer ;
    declare j                           integer ;
```

```
j=0;
final_ptr=init_ptr;
IF final_ptr<>0 THEN                        ****
    DO;
         j=1;
    END;
```

B.2 Extract from disassembled code from the PROM

The lines marked **** contain the compilation error. The compiler attempts to load both zero and the most significant half of the pointer into DI; it then compares DI with itself.

```
0003              MOV j,#0000; word op
0009              LES DX,init_ptr
000C              MOV final_ptr,DX
000F              MOV final_ptr+00002,ES
0012              MOV SI,#0000                       ****
0015              MOV DI,SI                          ****
0017              MOV DI,ES                          ****
0019              CMP DI,DI                          ****
001B              JNE @000001; short jump
001D              CMP SI,DX
001F @000001:     JE @000002; short jump
0021              MOV j,#0001; word op
0027 @000002:     ...
```

B.3 Extract from modelling of PL/M-86 code in MALPAS Intermediate Language

```
j := 0;
final_ptr := init_ptr;
IF final_ptr <> 0 THEN
    BLOCK
        j := 1
    ENDBLOCK
ENDIF;
```

B.4 Extract from post-processed semantic analysis of PL/M-86

Following translation into MALPAS IL, semantic analysis, and post-processing, it can be seen that the PL/M-86 code correctly performs a comparison of the pointer against zero.

```
PROC pointer_comp_start_end;
IF NOT(init_ptr <> 0)
THEN MAP
   final_ptr := init_ptr;
   j := 0;
ENDMAP
ELSIF init_ptr <> 0
THEN MAP
   final_ptr := init_ptr;
   j := 1;
ENDMAP ENDIF
ENDPROC;
```

B.5 Extract from modelling of ASM-86 code (from the PROM) in MALPAS IL

```
final_ptr := pointer_value
            (65536*highword (final_ptr) + (dx));
                                     [ MOV FINAL_PTR__DWLOW,DX ]
final_ptr := pointer_value (lowword (final_ptr) + 65536*(es));
                                     [ MOV FINAL_PTR__DWHIGH,ES ]
si := 0;                                        [ MOV SI,0 ]
di := si;                                       [ MOV DI,SI ]
di := es;                                       [ MOV DI,ES ]
cf := dword_value (65536*di + si )
     < dword_value (65536*di + dx);
of := undefined;
sf := undefined;
zf := dword_value (65536*di + si )
     = dword_value (65536*di + dx);

                                                    [CMP DI,DI]
                                                    [JNE @00001]
                                                    [CMP SI,DX]
IF zf THEN GOTO 150 [@00002]        [ @00001:   JE @00002 ]

    ENDIF;
    ...
```

B.6 Extract from post-processed semantic analysis of ASM-86

Following translation into MALPAS Intermediate language, semantic analysis, and post-processing, it is seen that the ASM-86 code compares the pointer against a double word composed of the most significant part of the pointer combined with a least significant value of zero.

```
PROC pointer_comp_start_end;
IF dword_value(init_ptr)
   <> dword_value(65536 * highword(init_ptr))
THEN MAP
   final_ptr := (init_ptr);
   j := 1;
ENDMAP
ELSIF dword_value(init_ptr)
     = dword_value(65536 * highword(init_ptr))
THEN MAP
   final_ptr := (init_ptr);
   j := 0;
ENDMAP ENDIF;
ENDPROC;
```

B.7 Extract from file offered for compliance analysis

The PL/M-86 semantic results are processed into a post-condition (i.e. a "specification") to be compared with the ASM-86 results (presented as the "code body").

```
PROCSPEC pointer_comp_start_end(
 IN init_ptr : pointer
 OUT final_ptr : pointer
 OUT j : integer )

PRE (true)
POST  ((NOT('init_ptr <> 0))
   -->((final_ptr EQ ('init_ptr)) AND (j = (0))))
 AND  (('init_ptr <> 0)
   -->((final_ptr EQ ('init_ptr)) AND (j = (1))));

PROC pointer_comp_start_end;

IF (init_ptr) <> (65536 * highword(init_ptr))
THEN MAP
   final_ptr := (init_ptr);
   j := 1;
ENDMAP
ELSIF (init_ptr) = (65536 * highword(init_ptr))
THEN MAP
   final_ptr := (init_ptr);
   j := 0;
ENDMAP ENDIF;
ENDPROC;
```

B.8 Threat obtained from compliance analysis

In this case a threat was obtained which does not reduce to 'false'. Precisely
what has gone wrong is not immediately apparent from inspection of the
threat alone, but close examination of the preceding stages of the analysis
soon reveals where the problem lies.

```
MALPAS Release 6.0.2 (2Mb)
19-AUG-1993 10:36 - Compliance Analysis

 Non-false threats from source-code comparison of: ima_err

pointer_comp_start_end

_threat := init_ptr = 65536*selectorof(init_ptr)
          AND NOT(nil = init_ptr)
```

Uses and misuses of formal methods in computer security

C.P. Pfleeger

Trusted Information Systems (UK) Limited, Kingston-upon-Thames

Abstract Formal methods are used in computer security to achieve high assurance of correct design and implementation of computer software. This use has been concentrated at the higher levels of assurance, and thus formal methods have come to be viewed as equivalent to high assurance. Both aspects of the assertion that formal methods are equivalent to high assurance have been misunderstood and misused. This paper will survey the uses of formal methods—especially in the U.S.—in computer security and recommend how formal methods might appropriately be used.

1 Uses of formal methods

Formal methods have been used in the U.S. predominantly in systems for the U.S. Department of Defense (DoD). The use of formal methods has heavily influenced the U.S. standards for evaluation of the trustworthiness of computer systems, the TCSEC [1]. However, large-scale use of formal methods was seen as at or beyond the state of the art as the TCSEC was being written.

1.1 Evaluation

1.1.1 Pre-TCSEC

In 1978 the U.S. National Bureau of Standards (NBS, now known as the National Institute of Standards and Technology, NIST) held a workshop on evaluation of trustworthiness of computer systems. Prior to that workshop there had been informal discussion of evaluation criteria and the drafting of several versions of evaluation requirements. The purpose of the workshop was to review and obtain comments on the then-current draft of evaluation requirements.

The evaluation criteria reviewed had ten levels, numbered 0 through 9, divided roughly into four groupings:

- **Level 0: Null**   This level was expected to entail no use of formal methods. No formal design specification would be required, and traditional development and testing techniques would be employed.

- **Level 1-3: Formal Design Specification**   This level would require the presentation of a design specification expressed using a formal (rigorously-defined) syntax. There would be no expectation for a formal proof of correctness of the design, nor any formal demonstration of the correspondence between the implementation and its specification.

- **Level 4-6: Proven Specification and Verifiable Implementation** This level would require that the system be developed using provable techniques, i.e., techniques from which the correspondence of steps throughout development could be performed, but was not required to be.

- **Level 7-9: Proven Design and Verified Implementation** The system would be developed using techniques for ensuring that the code accurately reflects its design and specification.

It is interesting that the use of formal (used in the sense of "rigorous") methods appears quite early in these requirements. However, this approach did not match the current state of practice, either for general commercial products, or for many special-purpose, bespoke product procurements.

1.1.2   TCSEC

Following that workshop, subsequent drafts of requirements led to a version similar to the final TCSEC, but again different in the formal methods requirements. That version contained classes D, C1, C2, B1, B2, B3, A1 and A2, whereas the final version of the TCSEC went only to class A1. There was consideration of a class A0, requiring presentation of a formal (stated in a language with precisely defined syntax and semantics) top level specification, but *not* requiring verification of that specification. Classes A2 (and A3, ...) were recognised as being beyond the state of the art, but would require formal (i.e., mechanical) analysis of source code, object code, compilers, loaders, hardware, and firmware.

The actual TCSEC classes are D, C1, C2, B1, B2, B3, and A1. A formal presentation of a security policy model is required at class B2, and that model must be proven to be adequate to enforce the security objectives. Informal demonstrations are used to argue that the product's design is consistent with this model.

The next significant change involving use of formal methods occurs at class A1, in which the requirements include:

- formal description of the security policy model, (same requirement as class B2)

- proof that the model is sufficient to enforce its security policy, (same requirement as class B2)

- formal model of trusted computing base (TCB) policy proven consistent with its axioms,

- formal top level specification (FTLS) that accurately describes TCB exceptions, error messages, and effects,

- FTLS shown to be an accurate description of the TCB interface,

- FTLS shown to be consistent with the formal security policy model,

- use of formal methods in analysis for covert channels,

- identification of specific TCB protection mechanisms, and explanation to show that they satisfy the formal policy model,

- manual or other mapping of FTLS to TCB source code to provide evidence of correct implementation.

The increase in use of formal methods (and consequently in complexity of implementation) is extremely large between classes B3 and A1.

1.2 Formal methods in practice

Formal methods have been used in a number of U.S. projects, as shown in the examples in Table 1. The point shown in this table is that several projects involving use of formal methods were commenced in the mid- to late-1970s, mostly prototype or research projects. Transfer to commercial products began to occur during the mid-1980s but, for a variety of reasons, not all of those projects concluded successfully. In the absence of significant government funding, the commercial demand was not adequate to support blossoming of use of formal methods.

1.3 Problems with use of formal methods in the U.S.

Four problems have developed with respect to the use of formal methods on ordinary commercial activities in the U.S.

1.3.1 Funding

Projects involving use of formal methods in the U.S. have suffered from difficulties of funding. The schedules of projects involving formal methods have tended to slip. Realistic or accurate or acceptable estimates of the

Table 1. Examples of projects using formal methods in U.S.

Project	Begun	Status
MITRE 11/45 Kernel	1976	prototype
PSOS (Provably Secure Operating System)	1977	research project
Kernelised VM/370 (KVM/370)	1977	discontinued
KSOS (Kernelised Secure Operating Systems)	1977	limited use
AUTODIN II	1978	project terminated (for reasons other than use of formal methods)
UCLA Data Secure Unix	1980	prototype
SCOMP (Secure Communications Processor)	1983	evaluated A1, 1985
GEMSOS (Gemini Secure Operating System)	1986	? (in evaluation, at ? class: B3?
ASOS (Army Secure Operating System)	1990	delivered, certified (?)
DEC VAX Security Kernel	1990	project discontinued

actual cost of use of formal methods are difficult to obtain. The tendency, however, has been to underestimate the true costs, so that schedules and budgets both tend to be exceeded. This degree of inaccuracy has given use of formal methods a poor reputation.

A possible reason for the underestimates of difficulty is the relative paucity of sound, mature tools for performing formal analysis. In the absence of high quality tools, much analysis must be performed or assisted by hand, which is very time-consuming.

1.3.2 Evaluation methodology

The methods for evaluating use of formal methods, and the skills of individuals for that evaluation are also not well advanced. There is, of course, a vicious circle: until formal methods are heavily used, experience in their evaluation cannot be gained, and thus evaluations will be slower and more tedious, which leads to greater reluctance to use formal methods.

1.3.3 Demands: Size and complexity

The situations in which the use of formal (i.e., automated) methods could be of most value are those in which humans are incapable of coping with either the size or the complexity of the project. However, in order for the field of formal analysis to mature, it is desirable for the use of formal methods to progress from simple to more complex. Unfortunately, the demands for the use of formal methods have increased at a pace faster than the increase in the capability of their use.

1.3.4 Different perceptions: "Consumers" vs. "Producers"

In spite of advice to the contrary, "consumers" (persons acquiring systems for which formal methods have been used in the development) have unrealistically high expectations of the benefits of the uses of formal methods. In some instances, these consumers have been misled; in others, they have willingly assumed that use of formal methods necessarily implied correctness, in a variety of dimensions that formal methods could not address.

The demands of these same consumers have been inconsistent, however: while there was significant exploration of the use of formal methods in the 1970s, that exploration did not extend to significant support or demand for their use in the 1980s and beyond.

Producers (users of formal methods), on the other hand, have sometimes focused heavily on the mechanics of formal methods, concentrating on the syntax, not necessarily the semantics of the problem. In some cases, this focus on syntax has been unavoidable, as in the need to follow all potential violations of security indicated by inability of an analysis tool to demonstrate security. In some cases, use of formal methods has been concentrated on only one aspect of development of a product.

For all of these reasons, the use of formal methods has not been especially popular.

2 Where are formal methods today?

The general public—as well as some members of the technical community—misunderstand the benefits and limitations of use of formal methods. In order for formal methods to be used effectively, these misperceptions must be rectified.

2.1 Myths of formal methods

In an excellent analysis, Anthony Hall [2] listed seven myths associated with formal methods. His observations are especially relevant for this treatise. The myths he identified are shown in Table 2.

Table 2. Seven myths of formal methods (from Hall [2])

It is a *myth* that formal methods:

> can guarantee software is perfect
> work by proving software is correct
> are of benefit only for highly critical systems
> involve complex mathematics
> increase the cost of development
> are incomprehensible to clients
> are not used for *real* projects

2.2   Sources of myths

Where do these myths come from? There is no single source for these myths. Some problems arise from the producers (users) of formal methods, some from the consumers (purchasers of systems produced using formal methods) and some from other sources entirely.

2.2.1   The "guarantee software is perfect" myth

The origin of this myth is unclear, but it probably stems both from producers and consumers. Producers have had to express the utility of formal methods forcefully in order to have the new technology accepted. Some producers may have overstated the benefits of formal methods. In more cases, however, producers needed to express a rather complicated statement of what formal methods can do, to a rather unsophisticated audience, and it is likely that the hearers did not fully appreciate everything the speakers stated. Security adds to the problem of misperception because consumers perceive the security problem as binary: a system is either secure or it is not; there are no degrees of security.

2.2.2   The "proving correctness" myth

This myth doubtless arises from terminology. While formal verification tools—also known as "theorem provers"—do carry out the mathematical rigor of a proof, what they prove is often not quite what the consuming public perceives. In security, formal methods may demonstrate that a security policy model is consistent with a set of axioms, or that a formal representation of a high level specification is not inconsistent with a formal representation of a security policy model. Those demonstrations are a long way from having proven that the executing version of a program is "correct," for the public's normal use of the term "correct."

2.2.3 The "Increased cost of development" myth

This myth has some basis in actual development. It is true that formal methods have, on some projects, increased the costs or slowed the pace of development beyond their initial estimates. In part the delays have been due to the fact that use of formal methods is still an immature science, and the tools available for performing formal analysis are still evolving. Using immature tools will certainly slow down the pace of development. Unfortunately, the schedule slippages harm the credibility of proponents of formal methods, which reduces the desirability of use of formal methods in the future, which limits the ability of formal methods to mature.

2.2.4 The "incomprehensibility" myth

This myth is the fault in part of the discipline, in part of the nature and origins of the discipline, and in part of the proponents of use of formal methods. The discipline is a rigorous one, based on precision of terminology and presentation. The origins of the discipline are in mathematics and logic, in which many members of the general public do not have a sound background. Terminology and notation do seem arcane to the outside world, even though they may be quite easy to grasp by anyone adept at symbol manipulation. However, the presentation of a formal argument can be daunting to the uninitiated.

3   What should be done in the future?

The most important lessons to be learned from this review of the uses and misuses of formal methods are ones of moderation. First, it is important that formal methods be used as part of a full family of assurance techniques. Formal methods are not the only source of assurance. Furthermore, it is not good to equate use of formal methods with the highest level of assurance. If formal methods are used on only projects of highest assurance, their use will be quite limited, and the myths and misperceptions of the general public will remain. Nor has it been demonstrated that use of formal methods is the only or best way to achieve high assurance. Use of formal methods must be seen as part of a balanced set of assurance techniques.

Second, the formal methods technology, which is still relatively immature, must be allowed and encouraged to mature. Formal methods must be used on appropriately-sized projects involving reasonable estimates of their cost so that their reputation can be brought more in line with their actual capabilities. In order for use of formal methods to grow, their tools must progress from research items to stable, robust, commercial products. This progress will occur only as formal methods continue to be used.

Finally, it is important not to oversell the capabilities of formal methods. The general public has both excessively positive and excessively negative perceptions of the capabilities of formal methods. While some of these misperceptions are beyond the capabilities of the formal methods research communities to address, the acceptance of formal methods is actually hurt by unrealistic or hyperbolic claims for the benefits of use of formal methods. It is important to present frankly what are the benefits and limitations of the use of formal methods, so that the consuming public will not be surprised by the result obtained through their use.

Bibliography

1. U.S. Department of Defense. (1985). *Trusted Computer System Evaluation Criteria, document 5200.28-STD.*

2. Hall, A. (1990). Seven myths of formal methods. *IEEE Software.*

Logical methods in the formal verification of safety-critical software

C. Pulley and G.V. Conroy

*Department of Computation, University of Manchester Institute of
Science and Technology*

1   Introduction

This paper reports on some work that is currently in progress under the
SERC/DTI Advanced Technology Program Safety Critical Systems initiative (project number IED4/1/9035). The project covers research into the
design and checking process for data written in application specific languages and is a collaborative effort between UMIST, GEC ALSTHOM
Signaling Limited and Westinghouse Signals Limited.

The project concentrates on one particular language - the **Solid State
Interlocking (SSI)** data language. This language is used to define the functionality of safety-critical railway signaling interlocking installations based
on the SSI system (see [1]).

The purpose of the work reported in this paper is to describe a formal
procedure by which one may automatically verify the safety of a specific
SSI interlocking. Unlike other research work reported on elsewhere (see
e.g. [2] and [3]), we shall explicitly model the IFPs polling cycle within the
formalism to be presented shortly. This has the added benefit of enabling
us to more correctly model the SSIs behavior and to also formally verify
the safety of the IFP. Unlike [3], we are able to express and verify much
more general safety properties. Our work also differs from the above in that
we have concentrated on trying to improve the efficiency of our decision
procedure rather than by concentrating on the *modes* by which we may
verify certain types of safety property.

This paper contributes to current research on the verification of safety
properties for SSIs in a number of specific ways. Firstly, the use of our
modeling techniques (i.e. use of second-order monadic arithmetic with the
successor function) enables us to provide a precise logical classification of
the types of safety properties that may be verified, along with a means by
which communicating groups of interlockings may also be verified. The
explicit modeling of the IFP polling cycle is an additional contribution of
our work in this area. Finally, by relating our verification strategy to

garbage collection, we also contribute to this area by indicating how one may parallelise the verification of interlocking safety properties.

After giving an overview of the SSI system in Section 2, we go on to show how SSIs may be modeled using finite state machines in Section 3. In Section 4, we introduce a safety property language and then go on to give a logical classification of the types of safety properties that may be expressed within the language. This section is finished by demonstrating the relationship between garbage collecting and verifying the safety of an SSI. Section 5 discusses some ongoing research work, with Section 6 finishing the paper by detailing some relationships between the approaches adopted in this paper and those used elsewhere.

Finally, thanks go to Bob for allowing the authors to *plagiarize* the introductory sections of [4].

2 Overview of the SSI system

An SSI (**S**olid **S**tate **I**nterlocking) is a generic railway signal control system designed to replace the more expensive and less flexible electromechanical systems that are currently in use. Each SSI installation has the function of controlling the points and signals on one section of a railway network. Interlocking between these controls ensures that safety of train movements is maintained at all times.

2.1 Hardware

The overall concept of an SSI may be summarised as follows:

- A centralised safety-critical microcomputer system issues commands to, and receives status indications from, up to 63 safety-critical controllers (the **T**rackside **F**unctional **M**odules or TFMs) distributed along the railway and up to 15 other SSI interlockings (thus enabling train movements between different interlocking control authorities).

- Each TFM contains a dual-channel microcomputer system, for safety purposes, and it drives and monitors either one or two colour-light signals, or up to four sets of points, as well as monitoring the status of train detection systems, etc.

- In the interests of availability, the centralised equipment is fault-tolerant, with active redundant 1-out-of-2 and 2-out-of-3 hardware configurations used for non-safety and safety portions of the system respectively.

- Communication between the centralised and distributed parts of the system is achieved by securely-coded digital links, duplicated for availability, carried on screened twisted-pair copper cables for distances of up to 40km. In addition, 64 kbit/s channels in optical fiber PCM systems may be used for longer distance communication.

- The detailed functional requirements of each installation are defined by the applications data, which is compiled into about 60k bytes of object data, and stored in EPROMs in each channel of central equipment.

2.2 SSI data language

For every new application of SSI, data must be prepared for, among other things, the interlocking system. The prepared data is in the form of a series of source files, which are then passed through several compiler programs to create sets of object files used to program EPROMs for each type of hardware module. Once these data EPROMs are installed in the hardware, the fixed interlocking program uses the SSI data to define the conditions to be applied at each stage of interlocking operation.

2.2.1 Fixed interlocking program

For the purposes of this paper the fixed interlocking program consists of the **Interlocking Functional Program** (IFP). The IFP is broken down into a series of processing stages.

First, there is the initialisation stage. Of these there are three possible start up modes:

Start up Modes 1, 2 and 3

For the purposes of this project we have taken start up mode 3 as our initialisation stage. The remaining two start up modes can be thought of as special cases of this start up mode.

The IFP then repeats a processing stage called the major cycle. A major cycle is further broken down into 64 minor cycles. Each minor cycle is essentially used to interact with a particular TFM. Minor cycles are themselves broken down into 5 message cycles. The processing activity of these message cycles are detailed below:

- **Processing of Incoming Telegrams** - an incoming telegram from a TFM and possibly another interlocking are processed on this message cycle.

- **Processing of Flag Data** - roughly a $\frac{1}{64}^{th}$ of the flag data is processed on this message cycle.

- **Processing of Outgoing Telegrams** - an outgoing telegram to a TFM and possibly another interlocking are processed on this message cycle.

- **Updating of Timers** - 2 signal timers, 4 track circuit timers and 1 elapsed timer are processed on this message cycle.

- **Processing of Panel Requests** - if time is left (a minor cycle may only take a maximum of 30 ms), then a panel request is processed on this message cycle.

Each major cycle exercises the whole of the flag data and input and output telegram data. All other data is accessed as and when needed.

2.2.2   SSI applications data

The SSI applications data consists of a series of SSI source data files. The SSI source data files include the logical conditions which must be applied at various stages of interlocking operation. These include:

- Input and output telegrams (defining the logical relationship between bits in the telegrams sent to, and received from, trackside controllers and signaling functions within the interlocking).

- Flag operations (logical conditions for releasing routes and for implementing user-defined functions).

- Panel requests (logic associated with the operator interface).

- Points free to move (conditions affecting point operation)

- Map data (a simplified map of the layout, used by the interlocking program to search through the layout).

The interlocking program accesses some of these files regularly, others as required (see the previous subsubsection).

3   Modeling the SSI system using finite state machines

In this section we describe how we intend to formally model our SSI as a finite state machine. Before doing this we note the following important details of the implementation that allow us to simplify our presentation:

- Any SSI can only communicate with at most 63 TFMs, at most 15 other interlockings and receive at most 1023 different panel requests (i.e. there is a finite range of possible inputs and outputs to an SSI).

- An SSI can only be in one of a finite number of internal memory states (i.e. finite range of states).

These two facts together suggest that an SSI may indeed be modeled by a finite state machine. The following simply formalises this intuition.

Definition 1. *Fix a nonempty finite set S of states, a finite set I of inputs, a finite set O of outputs, an accessibility relation $R^1 \subseteq S \times I \times S \times O$ and a nonempty set $I_0 \subseteq I$ of initial states. A finite state machine is then the tuple $(S, I, O, R, I_0)^2$.*

We may now model an SSI as a finite state machine as follows:

- I is taken to be the set of all possible tuples $(l, i_0, .., i_{78})$, where l is a partial map from the set $\{0, .., 63\}$ to the set of all possible panel requests for this SSI[3], and $i_0, .., i_{63}$ are incomming telegrams from the TFMs and $i_{64}, .., i_{78}$ are incoming telegrams from other SSIs.

- O is taken to be the set of all possible tuples $(o_0, .., o_{78})$, where $o_0, .., o_{63}$ are outgoing telegrams to the TFMs and $o_{64}, .., o_{78}$ are outgoing telegrams to other SSIs.

- S is taken to be the set of all possible internal memory states of the particular SSI.

- R is defined by the major cycle of the particular SSI. So we have that:

$$R(\sigma, i, \sigma', o)$$

holds precisely when our SSI is started in state σ with inputs i and, after one major cycle, is in state σ' and has produced outputs o.

- $I_0 \subseteq S$ is taken to be the set of valid mode 3 start up states for the particular SSI.

Our previous section and remarks ensure that these sets are indeed finite. For reasons of mathematical simplicity we shall find it more convenient to model our inputs and outputs as being part of our finite state machines state. Thus, given a finite state machine $M = (S, I, O, R, I_0)$, we

---

[1]Notice that R may here be nondeterministic. This is something we shall discuss briefly, later on in this paper.

[2]Note that we have intentionally not included any final states for our finite state machine.

[3]This partial map models the fact that an SSI may only process at most one incoming panel request per minor cycle.

shall in fact work with the finite state machine $M' = (I \times S \times O, \emptyset, \emptyset, R', I \times I_0 \times O)$, where R' is given by:

$$R'((i, \sigma, o), (i', \sigma', o')) \text{ iff } R(\sigma, i, \sigma', o)$$

Since such machines no longer have any need to refer to their input and output sets, we shall simply talk about the finite state machine $M' = (I \times S \times O, R', I \times I_0 \times O)$.

4   Formally reasoning about the safety of an SSI

Our aim is to formally verify that an SSI never allows a railway to enter an unsafe state. What exactly we shall mean by the terms *safe* and *unsafe* shall not be described here (this is the subject of some research work that is currently being carried out by Westinghouse Signals Limited). Instead we shall define a formal logical language that is sufficiently expressive enough to capture our intuitive notions of *safe* and *unsafe*. Using this safety property language, we may then go on to concentrate on the problem of how we may formally verify that our SSI satisfies a given property. We shall here be working in the main with the dual problem of that presented in [3]. It shall shortly be made clear that our approach is to perform a behavioral analysis of our SSI (see [2] and [3]).

Definition 2. *Given a finite state machine $M = (S, R, I_0)$[4], we define an M-stream $\sigma_0, \sigma_1, \ldots$ to be an infinite sequence of members of S such that:*

$$\sigma_0 \in I_0 \text{ and } \forall i \geq 0 \cdot R(\sigma_i, \sigma_{i+1})$$

For simplicity, we insist on dead end states as being modeled by repeating the state ad infinitum.

When we talk of wishing to perform a behavioral analysis of our SSI, we mean that we wish to constrain the possible M-streams of our SSI (where M is the finite state machine that models our specific SSI). So that we may formalise this further we introduce the following logical language that shall enable us to express various behavioral properties of our SSI.

Definition 3. *Let TL be the set of terms built from the constant term 0, the variables x_i ($i \in \mathbb{N}$) and the unary function s. We now define FL to be the set of first-order formulae built using the terms TL, the unary predicates p_i ($i \in \mathbb{N}$), the binary connective $\leq$, the propositional connectives $\neg$, $\vee$, $\wedge$ and $\implies$ and the first-order quantifiers $\forall$ and $\exists$.*

---

[4]Recall that in the previous section we made the assumption that all inputs and outputs are to be modelled as state items. Thus, finite state machines can be thought of as finite directed graphs with at least one distinguished node.

As is usual, we shall adopt the convention of allowing the symbols $p, q, r, ...$ (with or without indexing) to also denote unary predicate symbols.

The formulae in FL shall be used to specify our required behavioral properties. The following definition describes how this may be done:

Definition 4. *Fix a finite state machine M (with $S \subseteq \mathbb{P}(\mathbb{N})$) an M-sequence $\sigma : \mathbb{N} \longrightarrow S$ and an assignment function $A : \{x_i | i \in \mathbb{N}\} \longrightarrow \mathbb{N}$. We define an interpretation function $i_A : TL \longrightarrow \mathbb{N}$ by structural induction on TL as follows:*

$$i_A(0) = 0$$
$$i_A(x_i) = A(x_i)$$
$$i_A(s(t)) = i_A(t) + 1$$

and the M-satisfaction relation $\sigma, i_A \models_M \theta$ by structural induction on $\theta \in FL$ as follows:

$$\sigma, i_A \models_M t_1 \leq t_2 \text{ iff } i_A(t_1) \leq i_A(t_2)$$
$$\sigma, i_A \models_M p_i(t) \text{ iff } i \in \sigma(i_A(t))$$
$$\sigma, i_A \models_M \neg\varphi \text{ iff } \sigma, i_A \not\models_M \varphi$$
$$\sigma, i_A \models_M \varphi \wedge \psi \text{ iff } \sigma, i_A \models_M \varphi \text{ and } \sigma, i_A \models_M \psi$$
$$\sigma, i_A \models_M \varphi \vee \psi \text{ iff } \sigma, i_A \models_M \varphi \text{ or } \sigma, i_A \models_M \psi$$
$$\sigma, i_A \models_M \varphi \Longrightarrow \psi \text{ iff } \sigma, i_A \models_M \varphi \text{ implies } \sigma, i_A \models_M \psi$$
$$\sigma, i_A \models_M \forall x_l \cdot \varphi \text{ iff for all } j \simeq_{x_l} i_A, \sigma, j \models_M \varphi$$
$$\sigma, i_A \models_M \exists x_l \cdot \varphi \text{ iff there exists } j \simeq_{x_l} i_A \text{ such that } \sigma, j \models_M \varphi$$

where $j \simeq_{x_l} i_A$ means that the interpretation function $j : TL \longrightarrow \mathbb{N}$ satisfies the following condition:

$$\forall m \in \mathbb{N} \cdot m \neq l \text{ implies } j(x_m) = i_A(x_m)$$

M-validity is now defined using the above definition of M-satisfiability as usual by:

$$\models_M \theta$$
$$\text{iff}$$

$\forall M$-sequence σ, assignment function $A : \{x_i | i \in \mathbb{N}\} \longrightarrow \mathbb{N} \cdot \sigma, i_A \models_M \theta$

The intuition behind the above definition is that the unary predicates p_i $(i \in \mathbb{N})$ enable us to assert simple (i.e. propositional) statements about particular states of our given finite state machine M (hence the reason why a state consists of the indexes of those unary predicates that are true at the given state). The terms in TL enable us to specify particular points, within our state sequences (e.g. an initial state, a next state or an arbitrary state), at which our unary predicates are to hold. The remaining logical connectives enable us to assert statements constraining the way a proposed M-sequence may develop.

The following proposition (due originally to [5]) provides one with a means by which the majority of our first-order quantifiers may be eliminated:

Proposition 5. *Fix a finite state machine $M = (S, R, I_0)$ with $S \subseteq \mathbb{P}(\mathbb{N})$. Then for any $\theta \in FL$, there is a $\theta' \in FL$ such that,*

$$\models_M \theta \quad \text{iff} \quad \models_M \theta'$$

and what is more, θ' has the following format:

$$\sigma_0 \wedge \forall t \cdot \rho_{t,s(t)} \Longrightarrow \forall x \cdot \tau_x$$

where σ_0, $\rho_{t,s(t)}$ and τ_x are quantifier free formulae that do not contain the relation symbol $\leq$ and only contain the terms indicated by their indexes.

Proof. By [6] we have a θ'' such that,

$$\models_M \theta \quad \text{iff} \quad \models_M \theta''$$

with θ'' having the following format:

$$\sigma_0 \wedge \forall t \cdot \rho_{t,s(t)} \Longrightarrow (\exists y \forall x \cdot y \leq x \Longrightarrow \tau_x)$$

where σ_0, $\rho_{t,s(t)}$ and τ_x are quantifier free formulae that do not contain the relation symbol $\leq$ and only contain the terms indicated by their indexes. Again by [6],

$$\neg\theta'' \text{ is } M\text{-satisfiable}$$
$$\text{iff}$$
$$\neg\theta'' \text{ is } M\text{-satisfiable by an ultimately periodic } M\text{-sequence}$$
$$\text{of phase } \phi \text{ and period } p$$

What is more we have that $\phi + p \leq 2^{\text{number of distinct unary predicates in } \neg\theta}$ Thus, since we have an ultimately periodic M-sequence (see [6]), we have an $n \in \mathbb{N}$ such that,

$$\neg\theta'' \text{ is } M\text{-satisfiable}$$
$$\text{iff}$$
$$\sigma_0 \wedge \forall t \cdot \rho_{t,s(t)} \wedge (\exists x \cdot s^n(0) \leq x \wedge \tau_x) \text{ is } M\text{-satisfiable}$$

where $s^0(0) = 0$ and $s^{n+1}(0) = s(s^n(0))$. Now let $q_0,..,q_n$ be distinct unary predicates that do not appear anywhere in θ''. Then we have that,

$$\sigma_0 \wedge \forall t \cdot \rho_{t,s(t)} \wedge (\exists x \cdot s^n(0) \leq x \wedge \tau_x) \text{ is } M\text{-satisfiable}$$
$$\text{iff}$$
$$\sigma_0 \wedge \mu_0 \wedge (\forall t \cdot \rho_{t,s(t)} \wedge \lambda_{t,s(t)}) \wedge (\exists x \cdot q_n(x) \wedge \tau_x) \text{ is } M\text{-satisfiable}$$

where μ_0 is the formula;

$$q_0(0) \wedge \bigwedge_{i=1}^{n} \neg q_i(0)$$

and $\lambda_{t,s(t)}$ is the formula;

$$\bigwedge_{i=0}^{n-1} (q_i(t) \implies q_i(s(t)) \wedge q_{i+1}(s(t))) \wedge (q_n(t) \implies q_n(s(t)))$$

We now obtain our required result by taking θ' to be the formula,

$$\sigma_0 \wedge \mu_0 \wedge (\forall t \cdot \rho_{t,s(t)} \wedge \lambda_{t,s(t)}) \implies (\forall x \cdot q_n(x) \implies \tau_x)$$

$\square$

The previous proposition thus shows that if we are satisfied with FL as being expressive enough to capture all our required safety properties, then we need only express safety properties in the form:

$$\sigma_0 \wedge \forall t \cdot \rho_{t,s(t)} \implies \forall x \cdot \tau_x$$

In fact [6] demonstrates that a similar result holds even if we allow second-order quantifiers! But this more general result is beyond the needs of this paper.

Example 1. *Consider the following simple track layout consisting of a single signal (S), that may display one of two aspects (green(g) and red(r)), and a single track circuit (T), that may be in one of two states (occupied and unoccupied). The signal S is an entry signal to a number of routes;*

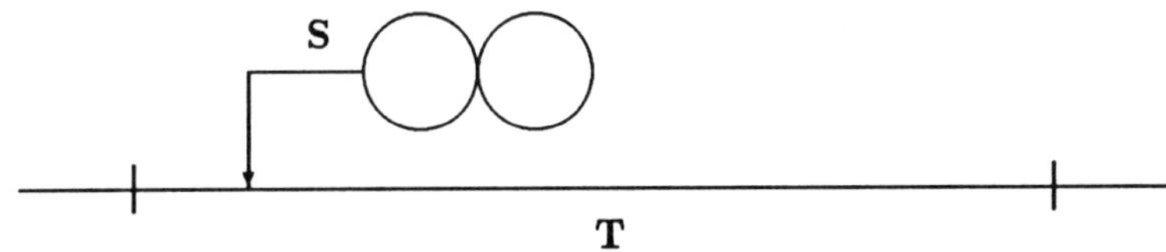

Figure 1.

By way of an example of the sort of safety properties that we have to verify, we shall consider a special case of approach locking. A signal may become approach locked when it displays a proceed aspect (i.e. green). Approach locking ensures that in the event of a signalman requesting (with a panel request Q) that a route (R), starting at our given signal S, be cleared, then any train that may not yet have cleared or reached the route is given sufficient time to either clear the route or stop at the signal S. The route may not be cleared until the signal has been cleared of approach locking. Release of approach locking may occur once a signal has changed to a red aspect and a timeout period has expired (for us this shall be a period of 180 major cycles). The following formulae describe the conditions for release of approach locking to occur;

1. $R_{set}(x) \wedge S_{aspect,g}(x) \wedge Q_{clear}(x) \implies$
   $S_{state,applock}(s(x)) \wedge S_{timer,180}(s(x)) \wedge S_{aspect,r}(s(x))$

2. $S_{state,applock}(x) \wedge \neg S_{timer,stopped}(x) \implies S_{state,applock}(s(x))$

3. $\bigwedge_{n=1}^{255}(S_{timer,n+1}(x) \implies S_{timer,n}(s(x)))$

4. $S_{timer,1}(x) \implies S_{timer,stopped}(s(x))$

5. $S_{timer,stopped}(x) \implies S_{timer,stopped}(s(x))$

Property 1 ensures that a signal shall become approach locked whenever the signal S is green, a route R is set from this signal and a panel request Q has been sent to clear the route. Under these conditions, the signal is to become approach locked, have its timer set for a 180 major cycle wait and its aspect set to red; Property 2 ensures that a signal maintains approach locked as long as the signals timer has not expired (ie. been stopped); Properties 3, 4 and 5 describe how a signals timer is updated over a given major cycle (these properties are contained within the logical description of the given SSI). Denoting property 1 and property 2 by the expressions $property_1(x, s(x))$ and $property_2(x, s(x))$ respectively we get the following safety property that is to be verified of our SSI (properties 3, 4 and 5 already being incorporated into the logical description of our SSI and thus do not need to be checked);

$$(\forall x \cdot property_1(x, s(x)) \wedge property_2(x, s(x)))$$

Placing this property into the normal form as given by Proposition 5 gives us the following safety property that is to be verified (where P_1 and P_2 are two new unary predicates);

$$(\forall x \cdot (property_1(x, s(x)) \iff P_1(x)) \wedge$$
$$(property_2(x, s(x)) \iff P_2(x))) \implies$$
$$(\forall y \cdot P_1(y) \wedge P_2(y))$$

Using the decidability algorithm originally due to [5], we can now (at least in principle) solve our original problem of verifying that a given safety property holds of an SSI. The remainder of this section describes a simplification of this original algorithm.

[2] and [3] appear to be concerned with demonstrating that a safety property is M-valid, whereas we shall instead mainly work with the dual problem of M-satisfiability. Thus, we shall be interested in showing that an SSI can never evolve into a state that fails our given safety property. The remainder of this section shall demonstrate the connection between this problem and the one of *garbage collection*.

Definition 6. *Let Mem be a (nonempty) set of memory cells, $\emptyset \neq Root \subseteq Mem$ and $Link \subseteq Mem \times Mem$. Then for $m \in Mem$, we say that:*

$$m \quad \text{is garbage iff there is no} \quad r \in Root \quad \text{such that} \quad (r, m) \in Link^*$$

where $Link^*$ is the reflexive, transitive closure of the relation $Link$.

Lemma 7. *Fix a finite state machine $M = (S, R, I_0)$ and a state $\sigma_f \in S$. Then the following problems are equivalent:*

$$\sigma_f \quad \text{is not garbage}$$
$$\exists M\text{-stream} \quad \sigma_0, \sigma_1, ... \exists i \in \mathbb{N} \cdot \sigma_i = \sigma_f$$

Proof. Take $Mem = S$, Link $= R$ and Root $= I_0$

The previous lemma demonstrates that if we view unsafe states as being garbage, then garbage collection can be thought of as providing the basis of a decision procedure for showing that an SSI satisfies a given property. This lemma is nothing more than a statement of the well known connection between the problem of reachability and garbage collection. Unfortunately, as we shall see in the next section, the memory spaces over which we wish to garbage collect are *astronomically* large. So usual garbage collecting strategies such as mark and sweep, copy collection, etc. just can not in general be applied in a computationally efficient manner.

However, we may exploit techniques such as lazily evaluating our garbage collecting algorithms, in order to help improve on the efficiency of our garbage collecting strategies. In addition we may also exploit this connection with garbage collection to help parallelise our decision procedure. To further improve our computational efficiency, some other techniques are required. Some possibilities are discussed in the next section.

5   Further research

We have seen in the previous section that any safety property expressed by
a formula of FL may be placed in the form,

$$\sigma_0 \wedge \forall t \cdot \rho_{t,s(t)} \Longrightarrow \forall x \cdot \tau_x$$

with quantifier free formulae σ_0, $\rho_{t,s(t)}$ and τ_x containing no relation symbols $\leq$ and only containing terms given by their indexes.

Since we are here interested in the problem of M-satisfiability, we are
interested in demonstrating the existence of an M-sequence σ that M-
satisfies $\sigma_0 \wedge \forall t \cdot \rho_{t,s(t)}$ and that fails to M-satisfy $\forall x \cdot \tau_x$. Now if we
think of M as being a finite state machine that models a given SSI, then
assuming that all predicates that occur in our safety property relate to
simple assertions about M's state, we may think of $\sigma_0 \wedge \forall t \cdot \rho_{t,s(t)}$ as placing
further constraints upon the evolutionary behavior of M. We could even
think of $\sigma_0 \wedge \forall t \cdot \rho_{t,s(t)}$ as *pruning* some of the IFP's possible execution paths.
This is a particularly attractive way of thinking about these formulae, since
it suggests a means by which we may further improve the computational
efficiency of our verification problem.

Unfortunately, it is not possible in general to think of $\sigma_0 \wedge \forall t \cdot \rho_{t,s(t)}$ in
this manner, since it can contain predicates that do not make assertions
about the state of M but rather are more concerned with simulating the
behavior of the $\leq$ relation, repeated applications of s, etc.

Some ongoing research work is to classify the roles of these additional
predicates in an attempt to better understand this relationship between our
safety properties and the way they may constrain the evolutionary behavior
of an SSI.

Due to the *astronomical* size of our search spaces (typically they contain
more states than the number of electrons in the universe! - see [7]) some
care needs to be exercised in how we implement our decision procedures.
Thus, the obvious mark and sweep garbage collecting strategy proposed in
the previous section is in general too costly in terms of our time and space
resources. This leads one to consider some means by which we may break
our overall problem into a collection of simpler ones.

One approach that has attracted some interest by other researchers (see
[3] and [8]) is that of decomposition by the use of topological properties
of our finite state machines. One aspect in which we differ from their
work, is that we intend to exploit the structure of the IFP in order to
decompose our verification problem into simpler portions. The intuitive
idea is that it is easier to verify that a property holds for each of the
message cycles of an arbitrary minor and major cycle than to show that
the same property holds of an arbitrary major cycle. Since a major cycle
is a sequential composition of message cycles, one would like to have some

way of *sequentially* composing safety properties that hold over particular message cycles. The following proposition establishes this property for some restricted types of safety property. However, we first need a technical lemma.

Lemma 8. *Let $S \subseteq \mathbb{P}(\mathbb{N})$, $\theta_{x,s(x)}$ be a quantifier free formula that only contains the terms given by its index. Furthermore let $A, A' : \{x_i | i \in \mathbb{N}\} \longrightarrow \mathbb{N}$ and $\sigma, \sigma' : \mathbb{N} \longrightarrow S$ be such that;*

$$\sigma(i_A(x)) = \sigma'(i_{A'}(x)) \quad and \quad \sigma(i_A(x)+1) = \sigma'(i_{A'}(x)+1)$$

Then;

$$\sigma, i_A \models \theta_{x,s(x)} \quad iff \quad \sigma', i_{A'} \models \theta_{x,s(x)}$$

where $\models$ means that we work not with M-sequences for some finite state machine M, but with arbitrary sequences drawn from the set S.

Proof. Proof proceeds by an easy induction on $\theta_{x,s(x)} \in FL$.

Definition 9. *Let $\theta_{x,s(x)}$ be a quantifier free formula that only contains the terms given by its indexes. We say $\theta_{x,s(x)}$ is transitive if the relation,*

$$\{(\sigma(n), \sigma(n+1)) | n \in \mathbb{N} \wedge \sigma : \mathbb{N} \to \mathbb{P}(\mathbb{N}) \wedge \sigma \models \theta_{x,s(x)}[s^n(0)/x]\}$$

over $\mathbb{P}(\mathbb{N})$ is transitive.

Proposition 10. *Let $M = (I_0, S, R)$ and $M' = (I'_0, S, R')$ be finite state machines with $S \subseteq \mathbb{P}(\mathbb{N})$ and such that the following conditions hold;*

$$R(a, b) \ implies \ \exists M\text{-sequence } \mu, i \in \mathbb{N} \cdot \mu(i) = a \wedge \mu(i+1) = b$$
$$R'(a', b') \ implies \ \exists M'\text{-sequence } \nu, j \in \mathbb{N} \cdot \nu(j) = a' \wedge \nu(j+1) = b'$$

for any $a, a', b, b' \in S$[5]. Then for any transitive quantifier free formulae $\theta_{x,s(x)}$ that only contains the terms given by its indexes, we have that;

$$\models_M \forall x \cdot \theta_{x,s(x)} \quad and \quad \models_{M'} \forall x \cdot \theta_{x,s(x)} \quad implies \quad \models_{M;M'} \forall x \cdot \theta_{x,s(x)}$$

where $M;M' = (I_0, S, R \circ R')$ and $R \circ R'$ is forward relational composition[6].

---

[5] i.e. essentially there are no redundant paths in the finite state machines M and M'.

[6] i.e. $R \circ R'(x, y)$ iff $\exists z \cdot R(x, z) \wedge R'(z, y)$

Proof. Proof proceeds by showing that for each $n \in \mathbb{N}$;

$$\models_M \forall x \cdot \theta_{x,s(x)} \quad \text{and} \quad \models_{M'} \forall x \cdot \theta_{x,s(x)} \quad \text{implies} \quad \models_{M;M'} \theta_{x,s(x)}[s^n(0)/x]^7$$

Our required result then follows easily from this. Now let σ be an $M;M'$-sequence and $n \in \mathbb{N}$ then the following two cases hold;

1. $R(\sigma(n), \tau)$ holds

2. $R'(\tau, \sigma(n+1))$ holds

where $\tau \in S$
By our initial assumptions we have that;

- we have an M-sequence σ' and an $m \in \mathbb{N}$ such that,

$$\sigma'(m) = \sigma(n) \text{ and } \sigma'(m+1) = \tau$$

- we have an M'-sequence σ'' and an $m' \in \mathbb{N}$ such that,

$$\sigma''(m') = \tau \text{ and } \sigma''(m'+1) = \sigma(n+1)$$

So by the hypothesis,

$$\models_M \forall x \cdot \theta_{x,s(x)}$$

we have that,

$$\sigma' \models_M \theta_{x,s(x)}[s^m(0)/x]$$

holds. By a similar argument we get that;

$$\sigma'' \models_{M'} \theta_{x,s(x)}[s^{m'}(0)/x]$$

holds. Using our previous lemma and the fact that $\theta_{x,s(x)}$ is *transitive* we get our required result.

Exploiting this intuitive idea directly is further complicated by the theoretical possibility that we may enter an *unsafe* state at the end of one message cycle and then find ourselves in a *safe* state again several message cycles later! And since time has been quantised into major cycles, this behavior would not be immediately observable to us. However, identifying such points of execution in our code might in itself be a useful way of performing a preliminary analysis of where we are likely to evolve into an *unsafe* state.

---

[7]By this we mean that for any $M;M'$-sequence μ,

$$\mu, j \models_{M;M'} \theta_{x,s(x)} \text{ holds}$$
for every interpretation $j : TL \longrightarrow \mathbb{N}$ such that $j(x) = n$

Similarly for $\mu \models_{M;M'} \theta_{x,s(x)}[s^n(0)/x]$.

Within the theoretical framework of this paper, it is possible to model control authorities consisting of more than one SSI. This is possible since we have allowed the possibility that our finite state machines may be nondeterministic. However, such models drastically increase the complexity of our verification procedures, and place a greater degree of emphasis on ensuring that we have good decomposition techniques to help deal with these extra complexity issues.

6   Other related work

This paper has focused on the use of finite state machines and their properties to the modeling of a particular computing problem. The approach adopted in this paper is by no means the only one. Indeed there are a number of other perfectly valid ways of modeling the problems addressed in this paper. This section discusses some of these alternative approaches.

6.1   Work related to the underlying logical theory

As indicated in the body of this paper, the underlying logical theory that we have utilised is that of Monadic Second-Order Arithmetic, equipped with just the constant 0 and the successor function. However, this is not the only way of viewing the underlying logical theory. One could have equally well have used Linear Propositional Temporal Logic (see eg. [9]) or have concentrated more on the underlying Büchi automata models (see [5]). However, we have avoided both these approaches here, preferring instead to work with the apparently more expressive system of second-order monadic successor arithmetic and to utilise properties of its automata models. This means that the safety properties that we express are thus closer to formulae of the first-order predicate calculus than they might otherwise be. This is of use when we consider that the eventual implementation of our decision procedures shall be used by people who not necessarily familiar with temporal operators or automata theory.
These different viewpoints do have their benefits though. They provide us with resolution techniques for verifying our safety properties (see [10]) and with worst case complexity results concerning our decision procedures (see [7]).

6.2   Work related to the modeling of SSIs

[11], [2] and [3] are the main approaches adopted so far to modeling the problems addressed within this paper.
[3] suggests that one should verify the safety of an SSI by using an inductive proof (ie. safety-transitivity), whilst relying on decomposition techniques

to help break the overall problem into smaller problems (see [8]). Unfortunately, this work only appears to deal with universal safety properties and there appears to be a need to verify stronger safety properties than might otherwise be necessary.

[11] suggests using CCS to model our SSIs with the concurrency workbench providing the tool by which we may proof check that a given safety argument is indeed valid.

[2] uses HOL to code-up a program logic for an SSI geographic data language with the intention of using HOL tactics to try and automate some of the proof search strategy. This appears to be a much more promising approach than that taken in [11]. HOLs tactic language should hopefully help to reduce the need for significant prompts for help from a user.

Both [11] and [2] suffer from the lack of any formal model of the IFP. [11], [2] and [3] all concentrate more on the *modes* by which one may verify certain types of safety property.

Bibliography

1. Cribbens, A.H. (1987). Solid State Interlocking (SSI): an intergrated electronic signaling system for mainline railways. *IEE Proceedings*, *134(3)*, pp. 148-158.

2. Morley, M.J. (1993). Safety in railway signaling data: A behavioral analysis. *Presented at the HOL User Group Meeting '1993*.

3. Ingleby, M. and Mitchell, I. (1992). Proving safety of a railway signaling system incorporating geographic data. *Proceedings SAFECOMP '92*, Zurich, Switzerland.

4. Barnard, B. (1993). Tool support for an application-specific language. *Safety-Critical Systems Club Conference*.

5. Büchi, J.R. (1960). On a decision method in restricted second-order arithmetic. *Logic Methodology and Philosophy of Science, Proceedings of 1960 Stanford International Congress*, Stanford 1962, pp. 1-11.

6. Siefkes, D. (1970). Büchi's monadic second order successor arithmetic. *Lecture Notes in Mathematics, 120*, Springer-Verlag.

7. Sistla, A.P. and Clarke, E.M. (1985). The complexity of propositional linear temporal logics. *Journal of the Association for Computing Machinery, 32*, 733-749.

8. Ingleby, M. (1993). A Galois theory of local reasoning in control systems with compositionality. *IMA Conference Proceedings on "Mathematics in Dependable Systems"*, Royal Holloway, London, pp. 103-130.

9. Gough, G.D. (1984). Decision procedures for temporal logic. PhD Dissertation, Department of Computer Science, University of Manchester.

10. Fisher, M. (1992). A normal form for first-order temporal formulae. *Proceedings of the Eleventh International Conference on "Automated Deduction"*, Saratoga Springs, New York., *Lecture notes in Computer Science, Vol. 607.*

11. Morley, M.J. (1991). Modeling British Rail's interlocking logic: Geographic data correctness. *LFCS Report ECS-LFCS-91-186*, University of Edinburgh.

Understanding software test adequacy[1] - An Axiomatic and measurement theory approach

H. Zhu, P.A.V. Hall and J.H.R. May

Department of Computing, The Open University, Milton Keynes

Abstract Objective and sensitive measurement of software test adequacy is particularly important where software is required to be of very high dependability. There are many criteria proposed in the literature for the measurement of the thoroughness of testing provided by a test set. However,there is no clear consensus as to how test adequacy should be measured. Whether existing proposed measures actually capture the 'true' notion of adequacy is a matter of some controversy. Each proposed adequacy criterion appears to have its strengths and weaknesses. This paper does not propose new adequacy criteria. Instead, the mathematical foundations of test adequacy are investigated. A list of axioms are suggested to capture the intuitive fundamental properties of test adequacy. Thus, the building blocks of a measurement theory for software testing are proposed.

1 Motivation

An important aspect of testing is being able to decide when enough testing has been performed. It follows that a measure of the thoroughness of software testing is needed. It is preferable to avoid subjectivity in such measures, and so it should be based on clear and sound theoretical foundations. Objective and sensitive measurement of software test adequacy is particularly important where software is required to be of very high dependability. In such cases it is not possible to rely on our intuition regarding thoroughness of testing. Clearly, software with a high dependability requirement needs to be tested more thoroughly than software for which a lower dependability is allowed. However, the amount of extra testing needed can only be decided relative to a testing thoroughness measure.

There are many criteria proposed in the literature for the measurement of the thoroughness of testing provided by a test set. Such criteria are usu-

---

[1] The work is funded by Nuclear Electric, Plc., UK, and it is part of CONTESSE project of DTI Safety Critical Systems Programme. The paper is published with the permission of Nuclear Electric, Plc.

ally called 'test adequacy criteria'. Examples include the families of control flow and data flow coverage, the mutation adequacy, and measures from statistical testing such as estimates for reliability or probability of failure on demand. However, there is no clear consensus as to how test adequacy should be measured. Whether existing proposed measures actually capture the 'true' notion of adequacy is a matter of some controversy. Each proposed adequacy criterion appears to have its strengths and weaknesses. This paper does not propose new adequacy criteria. Instead, the mathematical foundations of test adequacy are investigated. A list of axioms are suggested to capture the intuitive fundamental properties of test adequacy. Thus, the building blocks of a measurement theory for software testing are proposed.

Initial work on axioms for test adequacy studied test adequacy criteria as predicates, which discern test sets as either completely adequate or inadequate [1-3]. However, this paper will follow an approach based on a modern mathematical theory of measurement [4-8], which is the study of the logical and philosophical concepts underlying measurement, as it applies in all sciences. According to the theory, measurement is considered as an assignment of numbers to properties of objects in the real world by means of an objective empirical operation. The theory requires that the numbers assigned to the objects represent the perceived relation between those objects as determined by the properties. In this case, the objects to be measured are software tests with respect to the property of test adequacy. Then, as will be demonstrated in the paper, measurement theory suggests a test adequacy measure on a continuous scale.

Axiomatization plays the fundamental role in the development of a measurement that defines the perceived relation between the objects. Moreover, we can also distinguish four clear-cut roles which can be played by axiomatization in the research of software testing as they did in physics and mathematics.

Firstly, it makes explicit the exact details of a concept or an argument that, before the axiomatization, was either incomplete or unclear. For example, as pointed out by Parrish and Zweben [9], the discovery of *the applicability of test data adequacy criteria* is an excellent example of the role that axiomatization has played. *Applicability* was first proposed as an axiom by Weyuker in 1986 [1], requiring a criterion to be applicable to any program in the sense of the existence of finite test sets which satisfy the criterion. This axiom leads to the redefinition and re-examination of the power of some data flow criteria proposed earlier by Rapps and Weyuker in 1985 [10] which were found not applicable [11].

It is the main purpose of this paper to explore the properties which characterize the abstract and general notion of software test adequacy. It must be understood that there are two facets of the relationship between the existing test data adequacy criteria and the abstract notion of test data adequacy. The properties of the abstract and general notion of test adequacy should be confirmed by most existing criteria which are found interesting and useful in software testing practice. However, the existing criteria should be considered as approximations to an ideal measurement[2] of the abstract notion; they may differ from the ideal measurement in one or more properties. It is through the search into the properties which distinguish one criterion from another and distinguish a concrete criterion from the abstract notion that axiomatic study leads to a better understanding of test adequacy. Without a general abstract notion of software test adequacy to which existing criteria can be considered as approximations, existing concrete criteria can only be considered as different measures of different aspects of testing. Thus, the comparison of one with another would be meaningless (just like the comparison of a measure of the strength of a person with a measure of the intelligence), whereas, the comparison of the test data adequacy criteria has always been an active research subject in software testing (see e.g. [12]).

Secondly, axiomatization helps to isolate, abstract and study more fully a class of mathematical structures that have recurred in many important contexts, often with quite different surface forms. For example, random testing and various partition testing methods were considered difficult to put into a common mathematical structure, but axioms were found to characterize both of them [13]. In search of a model to prove the consistency of such an axiom system, we will define in this paper a mathematical structure covering both of the criteria in section 3.3.

For readers who do not accept the existence of a general abstract notion of software test adequacy, the axiom system studied in this paper can at least be understood as an attempt for this purpose, i.e. to abstract a class of test adequacy criteria which are different in the surface forms and to study their common structure and properties.

Thirdly, axiomatization can provide the scientist with a compact way to summarize and represent an empirical situation, one that allows a systematic exploration of the implications of the postulates. Relationships between various properties of test data adequacy criteria have been investigated by Weyuker [1], Parrish and Zweben [3,9] and by Zhu and Hall [13,24,25]. For instance, it is proved that finite applicability can be derived from other more basic properties of test data adequacy criteria [13]. The ultimate objective of our axiomatic study of test data adequacy criteria is to establish the relationships between adequacy measures and software in-

---

[2]Notice that, here we do not assume the existence of an EFFECTIVE measurement (of the abstract notion of test adequacy), where "effective measurement" means a computable function from the specifications, programs and test sets to degrees of test adequacy. That is, the ideal measurement may or may not be effective

tegrity measures, such as reliability, though it is still unclear whether it is achievable.

Finally, the fourth role is the study of what can be axiomatized in particular ways. For instance, it is an important subject whether formal properties can be defined to distinguish one type of adequacy criteria from another when criteria are expressed in a particular form such as predicates or functions into continuum.

The paper is organized as follows. The next section will briefly review the fundamental concepts of the measurement theory. Based on the theory, a framework for the measurement of test data adequacy will be developed. The irregularity of existing test data adequacy criteria will be proved. Section 3 will further investigate the axioms of test data adequacy criteria. The axioms from our previous work will be reviewed. The measurement theory properties of the axiom system, such as uniqueness, will be investigated. The notion of test adequacy in a test context will be introduced in the study of regression testing. Context sensitivity will be proved to be the property which distinguishes random testing from partition testing. Section 4 is the conclusion of the paper. The possibilities for further research will also be discussed.

2 The measurement theory

2.1 Basic notions and notation

Measurement is the assignment of numbers to properties of objects or events in the real world by means of an objective empirical operation. The modern form of measurement theory is representational, i.e. numbers assigned to objects/events must represent the perceived relation between the properties of those objects/events. Usually, such a theory comprises three main sections: the description of an empirical relational system, a representation theorem, and a uniqueness condition (See e.g. [4-8]).

An empirical relational system $\mathbf{Q} = (Q, R)$ consists of a set Q of manifestations of the property or attribute and a family of relations $R = \{R_1, R_2, \cdots, R_n\}$ on Q. The family R of relations provides the basic properties of the manifestations under measure. They are usually expressed as axioms derived from the empirical knowledge of the real world. To assign numbers to the objects in the empirical system, a numerical system $\mathbf{N} = (N, P)$ must be developed where N is a set of real numbers and $P = \{P_1, P_2, \cdots, P_n\}$ is a family of relations on N. Given an empirical relational system $\mathbf{Q}$ and a numerical system $\mathbf{N}$, the measurement is an objective empirical operation $M : Q \to N$ which maps attribute manifestations into numbers. The numbers must be assigned to the objects in such a way that the relations between numbers also hold between manifestations. This property is the most fundamental requirement for a valid measurement. It must be proved for the operation M, and it is usually called *representation theorem* in the measurement theory literature. Formally,

Representation theorem

Let P_i be a relation on N corresponding to R_i; then if $q, \cdots, r \in Q$,

$$R_i(q, \cdots, r) \Leftrightarrow P_i(M(q), \cdots, M(r)), \text{ for all } i = 1, 2, \cdots, n.$$

In other words, M is a homomorphism from **Q** to **N**. The conditions under which the theorem holds are called the *representation conditions*. Given an empirical relational system **Q** and a numerical relation system **N**,there may exist more than one measurement operation which satisfies the representation conditions. In most cases, two measurement operations M and M' are related one to another by a transformation f such that $M' = f(M)$. It follows from the representation theorem that:

$$R_i(q, \cdots, r) \Leftrightarrow P_i(f[M(q)], \cdots, f[M(r)]), \text{ for all } i = 1, 2, \cdots, n.$$

All such functions f comprise the class of *admissible transformations*. Admissible transformations characterize the uniqueness of the measurement. Hence, the conditions for a function to be an admissible transformation are known as the *uniqueness conditions*. Let M be a measurement which satisfies the representation conditions, $\mathbf{S} = < \mathbf{Q}, \mathbf{N}, M >$ is usually called a *scale* of measurement.

The measurement theory not only puts stress on the logic consistency of the axioms, but also on the validation of the axioms against the real world. To explore the empirical knowledge, the measurement theory approach involves: (1) expression of knowledge in the form of a relation on the manifestations; (2) validation of the relation against intuition and/or the real world; (3) definition of the measurement as a mapping from the empirical relation to the real numbers; and then, (4) proof of the representation theorem and uniqueness theorem.

In the remainder of the paper, we will follow this rigorous approach to develop a framework for test data adequacy measurement.

2.2 Test data adequacy criteria (TDAC) as measurements

As we are applying the measurement theory to software testing, the manifestations to be measured are the tests of programs with respect to specifications. We define a software test to be a triple (s, p, c), where s is a specification, p a program, and c a set of test cases. The property or attribute is the adequacy of the test. The empirical relation is, then, "a test v is of greater or equal adequacy compared to test u", written $u \le v$.

The most basic axioms about this relation are reflexivity, transitivity and comparability. Formally,

Axiom 1.   (Reflexivity) For all tests u, $u \leq u$.

Axiom 2.   (Transitivity) For all tests u, v and w, $u \leq v, v \leq w \Rightarrow u \leq w$.

Axiom 3.   (Comparability) For all tests u and v, $(u \leq v) \vee (v \leq u)$.

Axiom 1 states that any test is as adequate as itself. Axiom 2 states that if test w is at least as adequate as v, and v is at least as adequate as u, then w is at least as adequate as u.

Axiom 3 states that any two tests can be compared of their adequacy, i.e. either test u is at least as adequate as v or v is at least as adequate as u. This axiom might be thought as less intuitively acceptable, yet there is no evidence against the axiom. The arguments which support the axiom are, firstly, test adequacy is comparable if two different tests are applied to the same program and the same specification, say $u = (s, p, c_1)$ and $v = (s, p, c_2)$. Now, consider the situation that during testing p on the test set c_1, bugs are found in the program, and hence the program p is modified to form p' and re-tested on c_1. In such cases, we do compare the adequacy of the test of the modified program with the adequacy of testing the original program on the same test set. It often happens in regression testing that the adequacy of the re-test of the modified program is compared with the adequacy of the original test to see if additional test cases are required. In other words, the adequacies of testing two different programs by the same test set are comparable. Similarly, testing against different specifications are comparable.

Secondly, imagine that a software company has two software test teams, which perform the testing on two different programs (of different specifications). It would be desirable for the manager of the company to compare the quality of the testing performed by the two teams. Obviously, the thoroughness of the test (i.e. test adequacy) is one of the most important aspects of test quality.

Another point which supports axiom 1 to 3 is that, as we will show in the next subsection, almost all the test data adequacy criteria proposed in the literatures satisfy the axioms. In other words, comparability is confirmed by the existing concrete adequacy criteria. Hence, it should be a valid property of the abstract notion of adequacy.

It is worth noting that, if the comparability axiom is eliminated from the axiom system, the "$\leq$" relation on software testing becomes a partial ordering. Measurement theory is still applicable. A method to derive measurements on partial orderings has been developed in [14].

In this paper, we will focus on the criteria which satisfy the comparability property. Therefore, the following discussion will be based on the assumption of axioms 1 to 3. We will consider all the tests $< s, p, c >$ such that s is in a given class S of specifications, p in a given class P of programs, and c is a subset of the universal data space D. In practice, S can be the specifications written in a particular specification language such as Z, or a particular type of specifications such as algebraic specifications. P can be the programs written in a particular programming language such as Ada or Modula-2. The universal data space consists of all the data which can be used as input to some program in P. Without loss of generality, we assume that S, P and D are countable sets.

There are two special tests which should be noticed here. The first is the empty test, that is the software is not tested at all. Such tests will be denoted by $(s, p, \emptyset)$. Empty tests should be distinguished from the testing of a program which does not need any input data. Such tests will be denoted by $(s, p, \{\lambda\})$, where $\{\lambda\}$ means the execution of the program without input data. The second type of special test is exhaustive testing, where a program is executed on all data in the universal data space. These will be denoted by (s, p, D). Obviously, empty tests are the most inadequate tests whilst the exhaustive tests are the most adequate tests. Therefore, we have the following axioms.

Axiom 4.   (Boundedness) For all s, p and c, $(s, p, \emptyset) \leq (s, p, c) \leq (s, p, D)$.

Notice that the axiom 1 to 3 define a total ordering on the tests and axiom 4 specifies that the ordering has a maximum and a minimum element. The cardinality of the set $\{(s, p, c) | s \in S, p \in P \text{ and } c \subseteq D\}$ implies that the empirical system is homomorphic to the numerical system $([0,1], \leq)$, where $[0,1]$ is the unit real interval and $\leq$ is the less than or equal to relation on numbers. Therefore, by the measurement theory, a measurement of software test data adequacy must be a homomorphism from tests to real numbers in the unit interval. This leads to the following definition of test data adequacy criteria.

Definition 1. *Given S a class of specifications, P a class of programs, and D a universal data space, a test data adequacy criterion M is a mapping from the Cartesian product of S, P and the power set $\mathcal{P}(D)$ of D into the real interval $[0,1]$, i.e.*

$$M : S \times P \times \mathcal{P}(D) - > [0, 1]$$

such that $M(s, p, \emptyset) = 0$, and $M(s, p, D) = 1$. $\square$

We write $M_{s,p}(c)$ to denote $M(s, p, c)$. If $M_{s,p}(c) = r$, we say that according to the adequacy criterion M, the test set c is of adequacy r for testing the program p with respect to the specification s, or p is r-adequately tested by c with respect to s according to M. Intuitively, the value $M_{s,p}(C)$ gives a measure of how adequate the test set c is in testing the program p with respect to the specification s. The greater the value is, the more adequate the test set is. In the extreme case, if the value is 0, the testing is completely unacceptable. If the value is 1, the test is perfect. For the sake of convenience, we will omit the subscription of s and/or p when there is no possibility of confusion.

2.3 Examples of test data adequacy criteria

There are many TDACs proposed in the literature and used in software testing practices. The following are some examples.

1. Statement coverage:

$$M_p(c) = \frac{\text{number of primitive statements in } p \text{ executed by data in } c}{\text{number of feasible primitive statements in } p}$$

where a primitive statement is feasible if there is an input datum which can cause the execution of the statement. Similarly, we define the feasibility of a branch and a path in a program.

2. Branch coverage:

$$M_p(c) = \frac{\text{number of branches in } p \text{ exercised by data in } c}{\text{number of feasible branches in } p}$$

3. Simple path coverage:

$$M_p(c) = \frac{\text{number of simple paths in } p \text{ exercised by data in } c}{\text{number of feasible simple paths in } p}$$

4. Data flow path coverage:

There are various data flow adequacy criteria in the literature (see e.g. [10,11,15]). Here, we take Rapps, Weyuker and Frankl's [11] definition-use data flow path coverage as an example.

$$M_p(c) = \frac{\text{number of definition-use data flow paths in } p \text{ exercised by data in } c}{\text{number of feasible definition-use data flow paths in } p}$$

Above TDACs are all structure coverage measures. The following are error/fault based measures [16].

5. Strong mutation adequacy [17,18]:

$$M_p(c) =$$

$$\frac{\text{number of mutants of } p \text{ killed (i.e. distinguished from } p) \text{ by data in } c}{\text{number of mutants of } p \text{ - number of equivalent mutants}}$$

6. Weak mutation adequacy: [19]

$$M_p(c) = \frac{\text{number of mutants of } p \text{ locally distinguished by data in } c}{\text{number of mutants of } p \text{ - number of equivalent mutants}}$$

The above are all program-based TDACs. An example of specification-based adequacy criteria is specification mutation adequacy.

7. Specification mutation adequacy: [20]

By specification mutation testing, mutation operations are applied to the specification of the software to generate mutants of the specification. Then, the program under test is executed on test cases to obtain the output of the program. The input and output are used to evaluate the mutants of the specification. If a mutant is falsified by the test cases in the evaluation, we say that it is killed by the test cases, otherwise it remains alive. The adequacy of a set of test cases can be measured by using the following formula.

$$M_{s,p}(c) = \frac{\text{number of mutants of } s \text{ killed by data in } c}{\text{number of mutants of } s \text{ - number of equivalent mutants}}$$

The test adequacy criteria given above are functions from programs and/or specifications into real numbers. Therefore, these criteria satisfy the comparability because the test adequacy are numbers which can always be compared by the "less than or equal to" relation. It is also obvious that all the TDACs given above satisfy the condition of Definition 1. Some of the criteria are slightly different from those used in practice. For example, statement coverage is often calculated on the basis of the number of primitive statements instead of the number of feasible primitive statements. Without the feasibility requirement, the adequacy measure does not satisfy the condition $M_{s,p}(D) = 1$ when the program p contains dead code.

2.4   Irregularity of test data adequacy measurements

One of the important subjects in the research of measurement is the regularity of the measurement. Let $\mathbf{Q}$ be the empirical relational system, $\mathbf{N}$ be the numerical relational system, and M be a measurement. The scale $< \mathbf{Q}, \mathbf{N}, M >$ is said to be regular, if for every scale $< Q, N, M' >$, there is an admissible transformation from M to M'. If for every measurement M, $< \mathbf{Q}, \mathbf{N}, M >$ is regular, we call the representation $\mathbf{Q} \rightarrow \mathbf{N}$ regular. The following theorem gives a characteristics of regular scales.

Theorem 2.1 (Roberts and Franke 1976) [4]

$< \mathbf{Q}, \mathbf{N}, M >$ *is regular if and only if for every other measurement M' from $\mathbf{Q}$ to $\mathbf{N}$, and for all a, b in Q, $M(a) = M(b)$ implies $M'(a) = M'(b)$.*

From this theorem we can prove that any system of TDAC which contains any two of the test data adequacy criteria defined in the previous subsection is irregular.

Theorem 2.2. (Irregularity Theorem)

A measurement system of test data adequacy is irregular if it contains at least two of the following test data adequacy criteria, where the definition of the test data adequacy criteria is as in Section 2.3.

1. Statement coverage;

2. Branch coverage;

3. Simple path coverage;

4. Definition/use data-flow path coverage;

5. Strong mutation adequacy;

6. Weak mutation adequacy;

7. Specification mutation adequacy.

Proof. Here we only give the proof of the irregularity of branch coverage. The proofs for other criteria are very similar.

Consider the following two simple programs p_1 and p_2:

$$p_1 : \text{if } x = 0 \text{ then } x := 1; \quad y := 2 \text{ else } x := -1; \quad y := -2 \text{ endif}$$

$$p_2 : \text{if } x = 0 \text{ then } x := 1; \quad y := 2 \text{ else } x := -1 \text{ endif}$$

Let test set $c = \{x = 0\}$. Then, by statement coverage, we have that $M_{P1}(c) = 1/2$, $M_{P2}(c) = 2/3$. By branch coverage, $M'_{P1}(c) = M'_{P2}(c) =$

1/2. Roberts-Franke's Theorem implies that there are no transformations from branch coverage to statement coverage. Similarly, we can prove that there are no admissible transformations from branch coverage to other TDACs. Hence, branch coverage is irregular. $\square$

Regularity has been observed for various measurements of the physical objects such as length, weight, temperature, etc. It is also assumed in the research of software metrics [21]. With the assumption of regularity, the relationship between the types of scales and the meaningfulness of statements concerning measurements have been studied.

The *scale type* of a regular measurement is characterized by the class of admissible transformations. For example, the scale type of a regular measurement system is called nominal, if the admissible transformations are one-one functions. The scale type is called ordinal, if the admissible transformations are monotonic increasing functions. If the admissible transformations are linear functions of the form $f(x) = \alpha x + \beta$, $\alpha > 0$, the scale type is called *linear interval*. A more strict scale type is ratio, which requires that the admissible transformations are linear functions of the form $f(x) = \alpha x$, $\alpha > 0$. The *absolute* scale type is the strictest scale type which only has one admissible transformation $f(x) = x$.

It is widely understood that certain kinds of numerical statements involving measurement representations are meaningless in the sense that their truth value is dependent on which particular representation or representations are being employed. For example, the following statement is meaningless because it can be true in the degrees of Fahrenheit, but false in centigrade at the same time.

"The temperature of one can of corn at this moment is twice as much as that of a second can".

However, the following statement is meaningful because its truth remains the same no matter whatever measures are being used, say, ounces, pounds, kilogrammes, etc.

"The weight of one can of corn at this moment is twice as much as that of a second can".

The reason for the difference between the two statements is that the scale type of temperature measurement is linear interval, whilst the scale type of weight measurement is ratio. In general, given an empirical system Q and a numerical system N, a statement involving the measurement representations is meaningful if its truth is invariant for all measurements $M: Q \rightarrow N$. In regular measurement systems, the meaningfulness of a statement can be formally defined to be that its truth remains unchanged under all admissible transformations of all the scales involved [4]. Within this general domain of regular measurement, invariance under scale type has been a major feature of the discussion of the meaningfulness of particular statistical qualities, hypotheses, and tests formulated in terms of numerical measures, see e.g. [7].

In the light of measurement theory and the irregularity of existing TDACs proved above, we are seriously questioning the meaningfulness of statistical comparison of test data adequacy criteria. Such comparisons often take one criterion as the standard and test a number of sample programs on some sample test sets which satisfy the criteria under comparison. The adequacy of each test set is then measured according to the standard criterion. The averages of the adequacies and even the sizes of the test sets for each criterion are then considered as the evidence that one criterion is better than another. Similar work can be found in the research in software metrics. The same question about their meaningfulness should be asked. Our conjecture is that other software metrics, such as complexity measures, are also irregular.

There are a number of possible explanations of the irregularity of TDACs. Firstly, the irregularity may be due to the fact that software tests are very complicated objects to measure. It may be one of the essential differences between the physical objects and the objects in information science. Hence, a rigorous approach to the measurement is not a trivial task. Secondly, the irregularity implies that not all the TDACs used in practice are valid, i.e. catching the intuition of adequacy. Such criteria are actually different approximations to the ideal measure of abstract notion of test adequacy. Here, it is worth noting that, the irregularity of the existing test adequacy criteria must be distinguished from the property of the ideal measures of the abstract notion of adequacy, whose regularity is yet unknown. This situation is similar to the measurement of the size (i.e. volume) of lunch boxes by the number of hamburgers which can be put into the lunch box and by the numbers of the cups of coffee that the lunch box can contain. As different approximations to the measure of the volume of lunch boxes, these two measures are not regular.

Therefore, it is necessary to find the "errors" of each criterion. To do so, in the next section we will propose more axioms of test data adequacy measurements and discuss the properties which distinguish two classes of test adequacy criteria, partition testing and random testing.

3 Axioms of test data adequacy measurement

This section gives more axioms about test data adequacy criteria and some properties which are derivable from the axioms. The axioms as well as the derived properties will be expressed in the form of relations on measurements instead of relations of tests. This does not lose generality, because we will consider the representation theorem as the most fundamental requirement of the validity of adequacy measurements.

3.1 General axioms

When details of the program, specification and test set are taken into account, more axioms of test adequacy can be proposed. For example, the following axioms are from our previous work, which only concern the test set [13].

Axiom 5. (Inadequacy of empty test) $M(s,p,\emptyset) = 0$, for all programs p and specifications s.

Axiom 6. (Adequacy of exhaustive testing) $M(s,p.D) = 1$, for all p, s.

The axioms 5 and 6 above are only repeat of Definition 1.

Axiom 7. (Monotonicity) $u \subseteq v \Rightarrow M(s,p,u) \leq M(s,p,v)$, for all p, s.

Axiom 8. (Subadditivity) $M(s,p,u \cup v) \leq M(s,p,u) + M(s,p,v)$, for all p, s.

Axiom 9. (Convergence) For all p, s,

$$t_1 \subseteq t_2 \subseteq \cdots \subseteq t_n \subseteq t_n \subseteq \cdots \Rightarrow \lim_{i \to \infty} (M(s,p,t_i)) = M(s,p,\bigcup_{i=1}^{\infty} t_i)$$

Readers are referred to [13] for details about the validation of the axioms against intuition. Because the TDACs given in the previous section satisfy the axioms, the axiom system is consistent. The following theorem gives the uniqueness condition of admissible transformations of the measurement system.

Theorem 3.1. (Uniqueness Theorem)

A function f on [0,1] is an admissible transformation of the measurement system defined by the axioms 6 to 9, if and only if it satisfies the following conditions:

(a) Monotonic increasing and continuous on [0,1];

(b) $f(0) = 0$;

(c) $f(1) = 1$;

(d) For all x, $y \in [0,1]$, $f(x+y) \leq f(x) + f(y)$, if $0 \leq x+y \leq 1$.

Proof.

($\Rightarrow$) Let f be a function satisfying the conditions, and M be a measurement which satisfies the axioms 6 to 9. Then, it is easy to check that $f(M)$ also satisfies the axiom 6~9.

($\Leftarrow$) Notice that the range of test data adequacy measurement is densitive on [0,1]. For example, the statement coverage can be any rational number n/m in [0,1]. Consider the program:

$$p: \text{ if } x = 0 \text{ then } x := 1; \quad x := 2; \cdots; x := n \text{ else } x := 1;$$

$$x := 2; \cdots; x := m - n \text{ endif}$$

Let test set $c = \{x = 0\}$. Then, the statement coverage $M_{s,p}(c) = n/m$. It is easy to prove that the conditions (a) to (d) are necessary. $\square$

From these axioms, a number of properties of test data adequacy criteria can be proved. Among such properties, the most important one is finite applicability. Intuitively, a test adequacy criterion is finitely applicable, if every reasonable adequacy (i.e. not perfect testing) can always be achieved by a finite test set.

Proposition 3.1. *If a test data adequacy criterion satisfies axioms 6 to 9, then it is finitely applicable, i.e., for all programs p and specifications s,*

$$0 \le r < 1 \Rightarrow \exists c \subseteq D(c \text{ is a finite set } \wedge M_{s,p}(c) \ge r)$$

Proof. See [13] $\square$

Therefore, the test data adequacy criteria given in Section 2.3 are all finitely applicable. In Section 3.3, we will prove that adequacy measures for both partition testing and random testing all satisfy these axioms.

3.2   Regression testing and the law of diminishing returns

The software testing process can be considered as repeated test-debug cycles. Suppose that software was developed, tested and released to the user. In the maintenance phase, an additional test was performed. How to measure the adequacy of the additional regression test? The adequacy must be considered in the context of its testing history. The context of a test set for a new cycle is the test set comprising all previous tests on the software. Formally,

Definition 3.1. *The adequacy of a test t in the context c, written $M_{s,p}(t|c)$, is defined as*

$$M_{s,p}(t|c) = M_{s,p}(t \cup c) - M_{s,p}(c)$$

From axioms 6 to 9, the following properties of test adequacy measure in context can be proved.

Proposition 3.2. *Let M be a test data adequacy criterion which satisfies axiom 6 to 9, then it has the following properties:*

1. $M(t|\emptyset) = M(t)$

2. $M(\emptyset|c) = 0$

3. $(v \supseteq u) \Rightarrow M(v|c) \geq M(u|c)$

4. $M(t|c \cup d) \leq M(t \cup d|c)$

5. $M(t|c) = M(t - c|c)$

6. $M(u|v) + M(v|u) \leq M(u \cup v)$

7. $t_1 \subseteq t_2 \subseteq \cdots \subseteq t_n \subseteq \cdots) \Rightarrow \lim_{n \to \infty} (M_{s,p}(t_n|c)) = M_{s,p}(\bigcup_{n=1}^{\infty} t_n|c)$

Proof. Straightforward. $\square$

Proposition 3.3.

1. An adequacy criterion M satisfies subadditivity if and only if it satisfies (6) above;

2. An adequacy criterion M satisfies convergence if and only if it satisfies (7) above.

Proof.

1. Only ($\Rightarrow$) needs to be proved. Let M satisfy (6), then,

$$2M(u \cup v) - (M(u) + M(v)) = M(u|v) + M(v|u) \leq M(u \cup v).$$

2. Straightforward. $\square$

An interesting property of software testing is the law of diminishing returns. Intuitively, the more a program has been tested, the less a test set can contribute in the context. Formally, this can be expressed by the following axiom.

Axiom 10. (Law of diminishing returns). For all programs p, specifications s, and test set t,

$$c \subseteq d \Rightarrow M_{s,p}(t|d) \leq M_{s,p}(t|c)$$

The law of diminishing returns is stronger than subadditivity.

Proposition 3.4. *Law of diminishing returns implies subadditivity.*

Proof. Let u, v be any test sets. Then, $M(u) = M(u|\ \emptyset) \geq M(u|v) = M(u \cup v) - M(v)$. □

All the TDACs given in Section 2.3. satisfy the law of diminishing returns. This will be proved to be a special case of a general result about partition testing and random testing in the next section.

3.3 Random testing and partition testing

This section will investigate the properties of random testing and partition testing. It will be shown that context sensitivity is an important feature which distinguishes adequacy measures for partition testing from those for random testing. Partition testing is proved to be context sensitive, whilst random testing is context insensitive.

A large number of software testing methods are partition testing methods. Generally speaking, a partition testing method generates a family of subsets of the data space D from the program under test and/or the specification of the software. The adequacy of a test set is then measured according to how many subsets in the family are covered by the test set. A test set covering a subset in the partition family means that they are not disjointed, i.e. the subset has at least one representative in the test set. Here, the family of the subsets in the partition is a multi-set of the subsets of the universal data space, and is not required to be non-overlapping. Formally, we have the following definition.

Definition 3.2. *A test data adequacy criterion M for partition testing is defined on two mappings π and μ. The mapping π generates a family $\pi(s,p)$ of subsets of data from the program under test and the specification s, μ is a function from $\pi(s,p)$ to real unit interval $[0,1]$ such that for all specifications s and programs p,*

$$\sum_{X \in \pi(s,p)} \mu(X) = 1$$

The adequacy measure is then defined by the following equation:

$$M_{s,p}(t) = \sum_{X \in T} \mu(X)$$

where $T = \{X | X \in \pi(s,p) \wedge (X \cap t \neq \emptyset)\}$. □

Intuitively, the mapping μ gives a weight to each sub-domain in π. This is a generalization of various coverage measures which give equal weight to the sub-domains. For example, statement coverage is a partition testing method. For each statement in the program, there is a subset of data in the family π which contains the input causing the execution of the statement. The mapping μ is a constant function that for all X in $\pi(s,p)$, $\mu(X) = 1/\parallel \pi(s,p) \parallel$, where $\parallel \pi \parallel$ is the size of the family π, i.e. the number of statements in the program. Mutation adequacy is also a partition testing method. For each mutant q of the program p, the subset in the family π is the set of data which kills the mutant q. The mapping μ is defined the same as statement coverage. Similarly, we can prove that all the other TDACs given in Section 2.3 are adequacy measures for partition testing methods. In the previous subsection, we have seen that they obey the law of diminishing returns. The following proposition states that it is true for all partition testing methods.

Proposition 3.5. *Partition test methods satisfy all the axioms 6 to 10.*

Proof. Here, we only give the proof of law of diminishing returns. The proof of others are similar.

Let t, u, v be any test sets, $u \subseteq v$. By the definition of partition testing methods, we have that

$$M(t|u) = \sum_{X \in T_u} \mu(X)$$

where $T_u = \{X \in \pi | (t \cap X \neq \emptyset) \wedge (u \cap X = \emptyset)\}$. By $u \subseteq v$, we have that $T_v \subseteq T_u$. Therefore, $M(t|u) \geq M(t|v)$. $\square$

In contrast to partition testing, random testing generates test sets randomly according to a probability distribution on the input space of program without dividing the input space into sub-domains [19]. Given a probability distribution function P, the adequacy of a test set c for random testing can be measured by the probability $P(c)$, i.e. the probability of the input falls in c. This measure can be considered as a degenerate case of partition testing where the family π of sub-domains is the set $\{\{x\}|x \in D\}$, and the μ function is the probability $P(\{x\})$. Therefore, random testing also satisfies the axioms 6 to 10.

The difference between random testing and partition testing is in the context sensitivity. Intuitively, an adequacy measure is said to be context free if the adequacy of a test set remains constant regardless of its context. Otherwise, the adequacy measure is called context sensitive. Formally,

Definition 3.3. *A test data adequacy criterion M is context free, if*

$$(t \cap u = \emptyset, t \cap v = \emptyset) \Rightarrow M_{s,p}(t|u) = M_{s,p}(t|v)$$

Otherwise, it is called context sensitive. $\square$

Context free adequacy measures have the following characteristics.

Theorem 3.2. *The following statements are equivalent.*

1. The test data adequacy criterion M is context free;

2. The test data adequacy criterion M has the property of additivity, i.e. for all test set u, v,

$$(u \cap v = \emptyset) \Rightarrow M_{s,p}(u \cup v) = M_{s,p}(u) + M_{s,p}(v)$$

3. According to the test data adequacy criterion M, every test set u has a fixed contribution to test a given program with respect to a given specification but regardless of the context of the testing, i.e.

$$\forall u \subseteq D \cdot \exists a_u \in [0,1] \cdot \forall c \subseteq D \cdot (u \cap c = \emptyset \Rightarrow M_{s,p}(u|c) = a_u)$$

In particular, every test datum has a fixed contribution in all contexts, i.e.

$$\forall (x \in D \cdot \exists a_x \in [0,1] \cdot \forall c \subseteq D \cdot (x \notin c \Rightarrow M_{s,p}(\{x\}|c) = a_x))$$

and $\forall u \subseteq D \cdot (M_{s,p}(u) = \sum_{x \in u} a_x)$.

Proof. By definition. $\square$

By theorem 3.2, we can prove that

Theorem 3.3.

1. Non-degenerate partition testing methods are context sensitive, where a partition testing method is non-degenerate, if there is a program p and a specification s such that there exists an $X \in \pi(s,p)$ such that $\| X \| > 1$ and $\mu(X) > 0$.

2. Random testing methods are context free.

Proof.

1. Let $A \in \pi$ such that $\mu(A) > 0$, and $\| A \| > 1$, say, $A = \{a,b\}$. Let $u = \emptyset$, $v = \{a\}$ and $t = \{b\}$. Then, $M(t|u) = \mu(A) > 0$. But, $M(t|v) = 0$. Hence, by definition, the adequacy measure is context sensitive.

2. Notice that probability distribution function P has additivity, i.e. for all x, y, $x \cap y = \emptyset \Rightarrow P(x \cup y) = P(x) + P(y)$. (See e.g. [23]) by Theorem 3.2, the adequacy measure is context free.

4 Conclusion and further research

This paper proposed a measurement theory approach to the measurement of software test data adequacy. Within this framework, axioms about test data adequacy criteria were investigated. The uniqueness theorem was proved. Then, the adequacy of regression testing, i.e. the adequacy of testing in context, and the adequacy of random testing and partition testing were studied.

Regularity is an important property of measurement, which directly related to the meaningfulness of a statement and statistic operation concerning measures. It is usually true in the measurements of physical objects. And it is usually assumed in the research in software metrics [14]. But, in this paper, we observed and proved the irregularity of existing test data adequacy criteria. This observation may have profound impacts on the research into the comparison of test data adequacy criteria, especially statistical comparisons.

The irregularity implies that existing test adequacy criteria are rough approximations to the measures of a general abstract notion of software test adequacy. Hence, further research into axioms or properties of test data adequacy measurement should be directed to find the "errors" of each criterion. The first step towards this direction, as we did in this paper, is to find the properties which distinguish one criterion from another.

In [13], axioms were proposed to distinguish structure coverage based criteria such as statement coverage and error/fault based criteria such as mutation adequacy. In this paper, we continued this thread by distinguishing two main classes of testing methods, i.e. random testing and partition testing, by using the notion of context sensitivity.

Preliminary work has also been done to investigate axioms and properties of test adequacy criteria which involves details about program structures and specifications [24,25]. We are continuing this thread of work in the framework of measurement theory reported in this paper.

Bibliography

1. Weyuker, E.J. (1986). Axiomatizing software test data adequacy. *IEEE trans. on. software engineering, SE-12, No. 12*, 1128-1138.

2. Weyuker, E.J. (1988). The evaluation of program-based software test data adequacy criteria. *C.ACM, 31, No. 6*, 668-675.

3. Parrish, A. and Zweben, S.H. (1991). Analysis and refinement of software test data adequacy properties. *IEEE trans. on Software Engineering, SE-17, No. 6*, 565-581.

4. Roberts, F.S. (1979). Measurement theory. *Encyclopedia of Mathematics and its applications*, *7*, Addison-Wesley.

5. Krantz, D.H., Luce, R.D., Suppes, P. and Tversky, A.(1971). Foundations of measurement. *Additive and polynomial representations*, *1*, Academic Press, New York.

6. Suppes, P., Krantz, D.H., Luce, R.D. and Tversky, A. (1989). Foundations of measurement. *Geometrical, threshold, and probabilistic representations*, *2*, Academic Press, San Diego.

7. Luce, R.D., Krantz, D.H., Suppes, P. and Tversky, A. (1990). Foundations of measurement. *Representation, axiomatization and invariance*, *3*, Academic Press, San Diego.

8. Finkelstein, L. and Leaning, M.S. (1984). A review of the fundamental concepts of measurement. *Measurement*, *2*, *No. 1*, 25 - 34.

9. Parrish, A.S. and Zweben, S.H. (1993). Clarifying some fundamental concepts in software testing. *IEEE Transactions on Software Engineering*, *19*, *No. 7*, 742 - 746.

10. Rapps, S. and Weyuker, E.J. (1985). Selecting software test data using data flow information. *IEEE Transaction on Software Engineering*, *SE_11*, *No. 4*, 367-375.

11. Frankl, P.G. and Weyuker, J.E. (1988). An applicable family of data flow testing criteria. *IEEE Transaction on Software Engineering*, *SE_14*, *No. 10*, 129-150 and 1483-1498.

12. Hamlet, D. (1989). Theoretical comparison of testing methods. *Proc. ACM SIGSOFT 3rd Symp. Software Testing, Analysis, and Verification.*, ACM press, pp. 28 - 37.

13. Zhu. H. and Hall, P.A.V. (1993). Test data adequacy measurement. *Software Engineering Journal*, *8*, *No. 1*, 21 - 29.

14. Fenton, N.E. (1992). When a software measure is not a measure. *Software Engineering Journal*, *7*, *No. 5*, 357 - 362.

15. Clarke, L.A., Podgurski, A., Richardson, D.J. and Zeil, S.J. (1989). A formal Evaluation of data flow path selection criteria. *IEEE Transactions on Software Engineering*, *15*, *No. 11*, 1318-1332.

16. Morell, L.J. (1990). A theory of fault-based testing. *IEEE Transactions on Software Engineering*, *16*, *No. 8*, 844-857.

17. DeMillo, R.A., Lipton, R.J.A. and Sayward, F.G. (1978). Hints on test data selection: Help for the practising programmer. *Computer*, *11*, *No. 4*, 34 - 41.

18. Budd, T.A. (1981). Mutation analysis: Ideas, examples, problems and prospects. *Computer Program Testing*, Editors: B. Chandrasekaran and S. Radicchi, North-Holland,

19. Howden, W.E. (1982). Weak mutation testing and completeness of test sets. *IEEE Transactions on Software Engineering*, *SE-8, No. 4*, 371-379.

20. Gopal, A. and Budd, T. (1983). Program testing by specification mutation. *Technical Report, TR 83-17*, University of Arizona.

21. Fenton, N.E. (1991). Software metrics - A rigorous approach. Chapman and Hall.

22. Duran, J.W. and Ntafos, S.C. (1984). An evaluation of random testing. *IEEE Transactions on software Engineering*, *SE-10, No. 4*, 438 - 444.

23. Billingsley, P. (1979). Probability and measure. John Wiley and Sons.

24. Zhu, H. and Hall, P.A.V. (1992). Testability of programs: Properties of programs related to test data adequacy criteria. *Technique Report, No. 92/05*, Department of Computing, The Open University.

25. Zhu, H. and Hall, P.A.V. (1992). Test data adequacy with respect to specifications and related properties. *Technique Report, No. 92/06*, Department of Computing, The Open University.